Critical acclaim for Ian Stewart

'The book consists of a series of
curious observations of the natu
more abstract, but in either c
intriguing, vivid and playful ...
greatest and most prolific popular ematics'

The Times

'Ian Stewart is a phenomenon in the world of popular maths with some 60 books under his belt'

Independent Saturday Magazine

'First-rate popular mathematics writing ... Stewart achieves what other popular mathematics writers merely strive for: an accurate, informative portrayal of contemporary mathematics without a single equation in sight' *Nature*

'Professor Stewart comes across as the Maths Teacher You Always Wanted, But Never Got: endlessly patient, sympathetic and interesting ... If you think you hate maths, let Prof Stewart convince you otherwise' *Daily Telegraph*

ALSO BY IAN STEWART

The Magical Maze
Does God Play Dice?
Figments of Reality
The Collapse of Chaos
From Here to Infinity
Fearful Symmetry
Another Fine Math You've Got Me Into
Game, Set, and Math
The Problems of Mathematics
Life's Other Secret

Ian Stewart was awarded the 1995 Michael Faraday Medal by the Royal Society for the year's most significant contribution to the public understanding of science. He graduated in mathematics from Cambridge University and is now Professor of Mathematics and Director of the Mathematics Awareness Centre at the University of Warwick. He is an active research mathematician and his present field is the effects of symmetry on dynamics, with applications to pattern formation and chaos theory. Much of his current writing is centred on popular science, and his books include *Does God Play Dice?* (an acclaimed bestseller which was translated into eleven languages) and *The Magical Maze.* Ian Stewart contributes to a wide range of newspapers and magazines in the UK, Europe and the USA.

NATURE'S NUMBERS

Discovering Order and Pattern in the Universe

IAN STEWART

PHOENIX

A PHOENIX PAPERBACK

First published in Great Britain
by Weidenfeld & Nicolson in 1995
This paperback edition published in 1996
by Phoenix, a division of Orion Books Ltd,
Orion House, 5 Upper St Martin's Lane,
London WC2H 9EA

Third impression 2001

Science Masters are published in association
with Basic Books,
a division of HarperCollins Publishers,
New York

A CIP catalogue record for this book
is available from the
British Library.

ISBN: 0 75380 530 8

Printed and bound in Great Britain by
The Guernsey Press Co. Ltd.,
Guernsey, Channel Islands

CONTENTS

PROLOGUE

THE VIRTUAL UNREALITY MACHINE

I have a dream.

I am surrounded by—nothing. Not empty space, for there is no space to be empty. Not blackness, for there is nothing to be black. Simply an absence, waiting to become a presence. I think commands: *let there be space*. But what kind of space? I have a choice: three-dimensional space, multidimensional space, even curved space.

I choose.

Another command, and the space is filled with an all-pervading fluid, which swirls in waves and vortices, here a placid swell, there a frothing, turbulent maelstrom.

I paint space blue, draw white streamlines in the fluid to bring out the flow patterns.

I place a small red sphere in the fluid. It hovers, unsupported, ignorant of the chaos around it, until I give the word. Then it slides off along a streamline. I compress myself to one hundredth of my size and will myself onto the surface of the sphere, to get a bird's-eye view of unfolding events. Every few seconds, I place a green marker in the flow to record the sphere's passing. If I touch a marker, it blossoms like a time-lapse film of a desert cactus when the rains come—and on every petal there are pictures, numbers, symbols. The sphere can also be made to blossom, and when it does, those pictures, numbers, and symbols change as it moves.

Dissatisfied with the march of its symbols, I nudge the sphere onto a different streamline, fine-tuning its position until I see the unmistakable traces of the singularity I am seeking. I snap my fingers, and the sphere extrapolates itself into its own future and reports back what it finds. Promising . . . Suddenly there is a whole cloud of red spheres, all being carried along by the fluid, like a shoal of fish that quickly spreads, swirling, putting out tendrils, flattening into sheets. Then more shoals of spheres join the game—gold, purple, brown, silver, pink . . . I am in danger of running out of colors. Multicolored sheets intersect in a complex geometric form. I freeze it, smooth it, paint it in stripes. I banish the spheres with a gesture. I call up markers, inspect their unfolded petals, pull some off and attach them to a translucent grid that has materialized like a landscape from thinning mist.

Yes!

I issue a new command. "Save. Title: A new chaotic phenomenon in the three-body problem. Date: today."

Space collapses back to nonexistent void. Then, the morning's research completed, I disengage from my Virtual Unreality Machine and head off in search of lunch.

*

This particular dream is very nearly fact. We already have Virtual Reality systems that simulate events in "normal" space. I call my dream Virtual Unreality because it simulates anything that can be created by the mathematician's fertile imagination. Most of the bits and pieces of the Virtual Unreality Machine exist already. There is computer-graphics software that can "fly" you through any chosen geometrical object, dynamical-systems software that can track the evolving state of any chosen equation, symbolic-algebra software that can take the pain out of the most horrendous calculations—and get them right. It is only a matter of time before mathematicians will be able to get inside their own creations.

But, wonderful though such technology may be, we do not need it to bring my dream to life. The dream is a reality now, present inside every mathematician's head. This is what mathematical creation feels like when you're doing it. I've resorted to a little poetic license: the objects that are found in the mathematician's world are generally distinguished by symbolic labels or names rather than colors. But those labels are as vivid as colors to those who inhabit that world. In fact, despite its colorful images, my dream is a pale shadow of the world of imagination that every mathematician inhabits—a world in which curved space, or space with more than three dimensions, is not only commonplace but inevitable. You probably find the images alien and strange, far removed from the algebraic symbolism that the word "mathematics" conjures up. Mathematicians are forced to resort to written symbols and pictures to describe their world – even to each other. But the symbols are no more that world than musical notation is music.

Over the centuries, the collective minds of mathematicians have created their own universe. I don't know where it is situated—I don't think that there *is* a "where" in any normal sense of the word—but I assure you that this mathematical universe seems real enough when you're in it. And, not despite its peculiarities but because of them, the mental universe of mathematics has provided human beings with many of their deepest insights into the world around them.

I am going to take you sightseeing in that mathematical universe. I am going to try to equip you with a mathematician's eyes. And by so doing, I shall do my best to change the way you view your own world.

CHAPTER 1

THE NATURAL ORDER

We live in a universe of patterns.

Every night the stars move in circles across the sky. The seasons cycle at yearly intervals. No two snowflakes are ever exactly the same, but they all have sixfold symmetry. Tigers and zebras are covered in patterns of stripes, leopards and hyenas are covered in patterns of spots. Intricate trains of waves march across the oceans; very similar trains of sand dunes march across the desert. Colored arcs of light adorn the sky in the form of rainbows, and a bright circular halo sometimes surrounds the moon on winter nights. Spherical drops of water fall from clouds.

Human mind and culture have developed a formal system of thought for recognizing, classifying, and exploiting patterns. We call it mathematics. By using mathematics to organize and systematize our ideas about patterns, we have discovered a great secret: nature's patterns are not just there to be admired, they are vital clues to the rules that govern natural processes. Four hundred years ago, the German astronomer Johannes Kepler wrote a small book, *The Six-Cornered Snowflake*, as a New Year's gift to his sponsor. In it he argued that snowflakes must be made by packing tiny identical units together. This was long before the theory that matter is made of atoms had become generally accepted. Kepler performed no experiments; he just thought very hard about various bits and pieces of common knowledge. His main evidence was the sixfold symmetry of snowflakes, which is a natural consequence of regular packing. If you place a large number of identical coins on a table and try to pack them as closely as possible, then you get a honeycomb arrangement, in which every coin—except those at the edges—is surrounded by six others, arranged in a perfect hexagon.

The regular nightly motion of the stars is also a clue, this time to the fact that the Earth rotates. Waves and dunes are clues to the rules that govern the flow of water, sand, and air. The tiger's stripes and the hyena's spots attest to mathematical regularities in biological growth and form. Rainbows tell us about the scattering of light, and indirectly confirm that raindrops are spheres. Lunar haloes are clues to the shape of ice crytals.

There is much beauty in nature's clues, and we can all recognize it without any mathematical training. There is beauty, too, in the mathematical stories that start from the clues and deduce the underlying rules and regularities, but it is a different kind of beauty, applying to ideas rather than things. Mathematics is to nature as Sherlock Holmes is to evidence. When presented with a cigar butt, the great fictional detective could deduce the age, profession, and financial state of its owner. His partner, Dr. Watson, who was not as sensitive to such matters, could only look on in baffled admiration, until the master revealed his chain of impeccable logic. When presented with the evidence of hexagonal snowflakes, mathematicians can deduce the atomic geometry of ice crystals. If you are a Watson, it is just as baffling a trick, but I want to show you what it is like if you are a Sherlock Holmes.

Patterns possess utility as well as beauty. Once we have learned to recognize a background pattern, exceptions suddenly stand out. The desert stands still, but the lion moves. Against the circling background of stars, a small number of stars that move quite differently beg to be singled out for special attention. The Greeks called them *planetes*, meaning "wanderer," a term retained in our word "planet." It took a lot longer to understand the patterns of planetary motion than it did to work out why stars seem to move in nightly circles. One difficulty is that we are inside the Solar System, moving along with it, and things that look simple from outside often look much more complicated from inside. The planets were clues to the rules behind gravity and motion.

We are still learning to recognize new kinds of pattern. Only within the last thirty years has humanity become explicitly aware of the two types of pattern now known as *fractals* and *chaos*. Fractals are geometric shapes that repeat their structure on ever-finer scales, and I will say a little about them toward the end of this chapter; chaos is a kind of apparent randomness whose origins are entirely deterministic, and I will say a lot about that in chapter 8. Nature "knew about" these patterns billions of years ago, for clouds are fractal and weather is chaotic. It took humanity a while to catch up.

The simplest mathematical objects are numbers, and the simplest of nature's patterns are numerical. The phases of the moon make a complete cycle from new moon to full moon and back again every twenty-eight days. The year is three hundred and sixty-five days long—roughly. People have two legs, cats have four, insects have six, and spiders have eight. Starfish have five arms (or ten, eleven, even seventeen, depending on the species). Clover normally has three leaves: the superstition that a four-leaf clover is lucky reflects a deep-seated belief that exceptions to patterns are special. A very curious pattern indeed occurs in the petals of flowers. In nearly all flowers, the number of petals is one of the numbers that occur in the strange sequence 3,5,8,13,21,34,55,89. For instance, lilies have three petals, buttercups have five, many delphiniums have eight, marigolds have thirteen, asters have twenty-one, and most daisies have thirty-four, fifty-five, or eighty-nine. You don't find any other numbers anything like as often. There is a definite pattern to those numbers, but one that takes a little digging out: each number is obtained by adding the previous two numbers together. For example, $3 + 5 = 8$, $5 + 8$

= 13, and so on. The same numbers can be found in the spiral patterns of seeds in the head of a sunflower. This particular pattern was noticed many centuries ago and has been widely studied ever since, but a really satisfactory explanation was not given until 1993. It is to be found in chapter 9.

Numerology is the easiest—and consequently the most dangerous—method for finding patterns. It is easy because anybody can do it, and dangerous for the same reason. The difficulty lies in distinguishing significant numerical patterns from accidental ones. Here's a case in point. Kepler was fascinated with mathematical patterns in nature, and he devoted much of his life to looking for them in the behavior of the planets. He devised a simple and tidy theory for the existence of precisely six planets (in his time only Mercury, Venus, Earth, Mars, Jupiter, and Saturn were known). He also discovered a very strange pattern relating the orbital period of a planet—the time it takes to go once around the Sun—to its distance from the Sun. Recall that the square of a number is what you get when you multiply it by itself: for example, the square of 4 is $4 \times 4 = 16$. Similarly, the cube is what you get when you multiply it by itself twice: for example, the cube of 4 is $4 \times 4 \times 4 = 64$. Kepler found that if you take the cube of the distance of any planet from the Sun and divide it by the square of its orbital period, you always get the same number. It was not an especially elegant number, but it was the same for all six planets.

Which of these numerological observations is the more significant? The verdict of posterity is that it is the second one, the complicated and rather arbitrary calculation with squares and cubes. This numerical pattern was one of the key steps toward Isaac Newton's theory of gravity, which has explained all sorts of puzzles about the motion of stars and planets. In contrast, Kepler's neat, tidy theory for the number of planets has been buried without trace. For a start, it *must* be wrong, because we now know of nine planets, not six. There could be even more, farther out from the Sun, and small enough and faint enough to be undetectable. But more important, we no longer expect to find a neat, tidy theory for the number of planets. We think that the Solar System condensed from a cloud of gas surrounding the Sun, and the number of planets presumably depended on the amount of matter in the gas cloud, how it was distributed, and how fast and in what directions it was moving. An equally plausible gas cloud could have given us eight planets, or eleven; the number is accidental, depending on the initial conditions of the gas cloud, rather than universal, reflecting a general law of nature.

The big problem with numerological pattern-seeking is that it generates millions of accidentals for each universal. Nor is it always obvious which is which. For example, there are three stars, roughly equally spaced and in a straight line, in the belt of the constellation Orion. Is that a clue to a significant law of nature? Here's a similar question. Io, Europa, and Ganymede are three of Jupiter's larger satellites. They orbit the planet in, respectively, 1.77, 3.55, and 7.16 days. Each of these numbers is almost exactly twice the previous one. Is *that* a significant pattern? Three stars in a row, in terms of position; three satellites "in a row" in terms of orbital period. Which pattern, if either, is an important clue? I'll leave you to think about that for the moment and return to it in the next chapter.

In addition to numerical patterns, there are geometric ones. In fact this book really ought to have been called *Nature's Numbers and Shapes.* I have two excuses. First, the title sounds better without the "and shapes." Second, mathematical shapes can always be reduced to numbers—which is how computers handle graphics. Each tiny dot in the picture is stored and manipulated as a pair of numbers: how far the dot is along the screen from right to left, and how far up it is from the bottom. These two numbers are called the coordinates of the dot. A general shape is a collection of dots, and can be represented as a list of pairs of numbers. However, it is often better to think of shapes *as* shapes, because that makes use of our powerful and intuitive visual capabilities, whereas complicated lists of numbers are best reserved for our weaker and more laborious symbolic abilities.

Until recently, the main shapes that appealed to mathematicians were very simple ones: triangles, squares, pentagons, hexagons, circles, ellipses, spirals, cubes, spheres, cones, and so on. All of these shapes can be found in nature, although some are far more common, or more evident, than others. The rainbow, for example, is a collection of circles, one for each color. We don't normally see the entire circle, just an arc; but rainbows seen from the air can be complete circles. You also see circles in the ripples on a pond, in the human eye, and on butterflies' wings.

Talking of ripples, the flow of fluids provides an inexhaustible supply of nature's patterns. There are waves of many different kinds—surging toward a beach in parallel ranks, spreading in a V-shape behind a moving boat, radiating outward from an underwater earthquake. Most waves are gregarious creatures, but some—such as the tidal bore that sweeps up a river as the energy of the incoming tide becomes confined to a tight channel—are solitary. There are swirling spiral whirlpools and tiny vortices. And there is the apparently structureless, random frothing of turbulent flow, one of the great enigmas of mathematics and physics. There are similar patterns in the atmosphere, too, the most dramatic being the vast spiral of a hurricane as seen by an orbiting astronaut.

There are also wave patterns on land. The most strikingly mathematical landscapes on Earth are to be found in the great *ergs*, or sand oceans, of the Arabian and Sahara deserts. Even when the wind blows steadily in a fixed direction, sand dunes form. The simplest pattern is that of transverse dunes, which—just like ocean waves—line up in parallel straight rows at right angles to the prevailing wind direction. Sometimes the rows themselves become wavy, in which case they are called barchanoid ridges; sometimes they break up into innumerable shield-shape barchan dunes. If the sand is slightly moist, and there is a little vegetation to bind it together, then you may find parabolic dunes—shaped like a U, with the rounded end pointing in the direction of the wind. These sometimes occur in clusters, and they resemble the teeth of a rake. If the wind direction is variable, other forms become possible. For example, clusters of star-shaped dunes can form, each having several irregular arms radiating from a central peak. They arrange themselves in a random pattern of spots.

Nature's love of stripes and spots extends into the animal kingdom, with tigers and leopards, zebras and giraffes. The shapes and patterns of animals and plants are a happy hunting ground for the mathematically minded. Why, for example, do so many shells form spirals? Why are starfish equipped with a symmetric set of arms? Why do many viruses assume regular geometric shapes, the most striking being that of an icosahedron—a regular solid formed from twenty equilateral triangles? Why are so many animals bilaterally symmetric? Why is that symmetry so often imperfect, disappearing when you look at the detail, such as the position of the human heart or the differences between the two hemispheres of the human brain? Why are most of us right-handed, but not all of us?

In addition to patterns of form, there are patterns of movement. In the human walk, the feet strike the ground in a regular rhythm: left-right-left-right-left-right. When a four-legged creature—a horse, say—walks, there is a more complex but equally rhythmic pattern. This prevalence of pattern in locomotion extends to the scuttling of insects, the flight of birds, the pulsations of jellyfish, and the wavelike movements of fish, worms, and snakes. The sidewinder, a desert snake, moves rather like a single coil of a helical spring, thrusting its body forward in a series of S-shaped curves, in an attempt to minimize its contact with the hot sand. And tiny bacteria propel themselves along using microscopic helical tails, which rotate rigidly, like a ship's screw.

Finally, there is another category of natural pattern—one that has captured human imagination only very recently, but dramatically. This comprises patterns that we have only just learned to recognize—patterns that exist where we thought everything was random and formless. For instance, think about the shape of a cloud. It is true that meteorologists classify clouds into several different morphological groups—cirrus, stratus, cumulus, and so on—but these are very general types of form, not recognizable geometric shapes of a conventional mathematical kind. You do not see spherical clouds, or cubical clouds, or icosahedral clouds. Clouds are wispy, formless, fuzzy clumps. Yet there is a very distinctive pattern to clouds, a kind of symmetry, which is closely related to the physics of cloud formation. Basically, it is this: you can't tell what size a cloud is by looking at it. If you look at an elephant, you *can* tell roughly how big it is: an elephant the size of a house would collapse under its own weight, and one the size of a mouse would have legs that are uselessly thick. Clouds are not like this at all. A large cloud seen from far away and a small cloud seen close up could equally plausibly have been the other way around. They will be different in shape, of course, but not in any manner that systematically depends on size.

This "scale independence" of the shapes of clouds has been verified experimentally for cloud patches whose sizes vary by a factor of a thousand. Cloud patches a kilometer across look just like cloud patches a thousand kilometers across. Again, this pattern is a clue. Clouds form when water undergoes a "phase transition" from vapor to liquid, and physicists have discovered that the same kind of scale invariance is associated with all phase transitions. Indeed, this *statistical self-similarity*, as it is called, extends to many other natural forms. A Swedish colleague who works on oil-field geology likes to show a slide of one of his friends standing up in a boat and leaning nonchalantly against a shelf of rock that comes up to about his armpit. The photo is entirely convincing, and it is clear that the boat must have been moored at the edge of a rocky gully about two meters deep. In fact, the rocky shelf is the side of a distant fjord, some thousand meters high. The main problem for the photographer was to get both the foreground figure and the distant landscape in convincing focus.

Nobody would try to play that kind of trick with an elephant.

However, you *can* play it with many of nature's shapes, including mountains, river networks, trees, and very possibly the way that matter is distributed throughout the entire universe. In the term made famous by the mathematician Benoit Mandelbrot, they are all fractals. A new science of irregularity—fractal geometry—has sprung up within the last fifteen years. I'm not going to say much about fractals, but the dynamic process that causes them, known as chaos, will be prominently featured.

Thanks to the development of new mathematical theories, these more elusive of nature's patterns are beginning to reveal their secrets. Already we are seeing a practical impact as well as an intellectual one. Our newfound understanding of nature's secret regularities is being used to steer artificial satellites to new destinations with far less fuel than anybody had thought possible, to help avoid wear on the wheels of locomotives and other rolling stock, to improve the effectiveness of heart pacemakers, to manage forests and fisheries, even to make more efficient dishwashers. But most important of all, it is giving us a deeper vision of the universe in which we live, and of our own place in it.

CHAPTER 2

WHAT MATHEMATICS IS FOR

We've now established the uncontroversial idea that nature is full of patterns. But what do we want to do with them? One thing we can do is sit back and admire them. Communing with nature does all of us good: it reminds us of what we are. Painting pictures, sculpting sculptures, and writing poems are valid and important ways to express our feelings about the world and about ourselves. The entrepreneur's instinct is to exploit the natural world. The engineer's instinct is to change it. The scientist's instinct is to try to understand it—to work out what's really going on. The mathematician's instinct is to structure that process of understanding by seeking generalities that cut across the obvious subdivisions. There is a little of all these instincts in all of us, and there is both good and bad in each instinct.

I want to show you what the mathematical instinct has done for human understanding, but first I want to touch upon the role of mathematics in human culture. Before you buy something, you usually have a fairly clear idea of what you want to do with it. If it is a freezer, then of course you want it to preserve food, but your thoughts go well beyond that. How much food will you need to store? Where will the freezer have to fit? It is not always a matter of utility; you may be thinking of buying a painting. You still ask yourself where you are going to put it, and whether the aesthetic appeal is worth the asking price. It is the same with mathematics—and any other intellectual worldview, be it scientific, political, or religious. Before you buy something, it is wise to decide what you want it for.

So what do we want to get out of mathematics?

Each of nature's patterns is a puzzle, nearly always a deep one. Mathematics is brilliant at helping us to solve puzzles. It is a more or less systematic way of digging out the rules and structures that lie behind some observed pattern or regularity, and then using those rules and structures to explain what's going on. Indeed, mathematics has developed alongside our understanding of nature, each reinforcing the other. I've mentioned Kepler's analysis of snowflakes, but his most famous discovery is the shape of planetary orbits. By performing a mathematical analysis of astronomical observations made by the contemporary Danish astronomer Tycho Brahe, Kepler was eventually driven to the conclusion that planets move in ellipses. The ellipse is an oval curve that was much studied by the ancient Greek geometers, but the ancient astronomers had preferred to use circles, or systems of circles, to describe orbits, so Kepler's scheme was a radical one at that time.

People interpret new discoveries in terms of what is important to them. The message astronomers received when they heard about Kepler's new idea was that neglected ideas from Greek geometry could help them solve the puzzle of predicting planetary motion. It took very little imagination for them to see that Kepler had made a huge step forward. All sorts of astronomical phenomena, such as eclipses, meteor showers, and comets, might yield to the same kind of mathematics. The message to mathematicians was quite different. It was that ellipses are really interesting curves. It took very little imagination for them to see that a general theory of curves would be even more interesting. Mathematicians could take the geometric rules that lead to ellipses and modify them to see what other kinds of curve resulted.

Similarly, when Isaac Newton made the epic discovery that the motion of an object is described by a mathematical relation between the forces that act on the body and the acceleration it experiences, mathematicians and physicists learned quite different lessons. However, before I can tell you what these lessons were I need to explain about acceleration. Acceleration is a subtle concept: it is not a fundamental quantity, such as length or mass; it is a rate of change. In fact, it is a "second order" rate of change—that is, a rate of change of a rate of change. The velocity of a body—the speed with which it moves in a given direction—is just a rate of change: it is the rate at which the body's distance from some chosen point changes. If a car moves at a steady speed of sixty miles per hour, its distance from its starting point changes by sixty miles every hour. Acceleration is the rate of change of velocity. If the car's velocity increases from sixty miles per hour to sixty-five miles per hour, it has accelerated by a definite amount. That amount depends not only on the initial and final speeds, but on how quickly the change takes place. If it takes an hour for the car to increase its speed by five miles per hour, the acceleration is very small; if it takes only ten seconds, the acceleration is much greater.

I don't want to go into the measurement of accelerations. My point here is more general: that acceleration is a rate of change of a rate of change. You can work out distances with a tape measure, but it is far harder to work out a rate of change of a rate of change of distance. This is why it took humanity a long time, and the genius of a Newton, to discover the law of motion. If the pattern had been an obvious feature of distances, we would have pinned motion down a lot earlier in our history.

In order to handle questions about rates of change, Newton—and independently the German mathematician Gottfried Leibniz—invented a new branch of mathematics, the calculus. It changed the face of the Earth—literally and metaphorically. But, again, the ideas sparked by this discovery were different for different people. The physicists went off looking for other laws of nature that could explain natural phenomena in terms of rates of change. They found them by the bucketful—heat, sound, light, fluid dynamics, elasticity, electricity, magnetism. The most esoteric modern theories of fundamental particles still use the same general kind of mathematics, though the interpretation—and to some extent the implicit world-view—is different. Be that as it may, the mathematicians found a totally different set of questions to ask. First of all, they spent a long time grappling with what "rate of change" really means. In order to work out the velocity of a moving object, you must measure where it is, find out where it moves to a very short interval of time later, and divide the distance moved by the time elapsed. However, if the body is accelerating, the result depends on the interval of time you use. Both the mathematicians and the physicists had the same intuition about how to deal with this problem: the interval of time you use should be as small as possible. Everything would be wonderful if you could just use an interval of zero, but unfortunately that won't work, because both the distance traveled and the time elapsed will be zero, and a rate of change of 0/0 is meaningless. The main problem with nonzero intervals is that whichever one you choose, there is always a smaller one that you could use instead to get a more accurate answer. What you would really like is to use the smallest possible nonzero interval of time—but there is no such thing, because given any nonzero number, the number half that size is also nonzero. Everything would work out fine if the interval could be made infinitely small—"infinitesimal." Unfortunately, there are difficult logical paradoxes associated with the idea of an infinitesimal; in particular, if we restrict ourselves to numbers in the usual sense of the word, there is no such thing. So for about two hundred years, humanity was in a very curious position as regards the calculus. The physicists were using it, with great success, to understand nature and to predict the way nature behaves; the mathematicians were worrying about what it really meant and how best to set it up so that it worked as a sound mathematical theory; and the philosophers were arguing that it was all nonsense. Everything got resolved eventually, but you can still find strong differences in attitude.

The story of calculus brings out two of the main things that mathematics is for: providing tools that let scientists calculate what nature is doing, and providing new questions for mathematicians to sort out to their own satisfaction. These are the external and internal aspects of mathematics, often referred to as applied and pure mathematics (I dislike both adjectives, and I dislike the implied separation even more). It might appear in this case that the physicists set the agenda: if the methods of calculus seem to be working, what does it matter *why* they work? You will hear the same sentiments expressed today by people who pride themselves on being pragmatists. I have no difficulty with the proposition that in many respects they are right. Engineers designing a bridge are entitled to use standard mathematical methods even if they don't know the detailed and often esoteric reasoning that justifies these methods. But I, for one, would feel uncomfortable driving across that bridge if I was aware that *nobody* knew what justified those methods. So, on a cultural level, it pays to have some people who worry about pragmatic methods and try to find out what really makes them tick. And that's one of the jobs that mathematicians do. They enjoy it, and the rest of humanity benefits from various kinds of spin-off, as we'll see.

In the short term, it made very little difference whether mathematicians were satisfied about the logical soundness of the calculus. But in the long run the new ideas that mathematicians got by worrying about these internal difficulties turned out to be very useful indeed to the outside world. In Newton's time, it was impossible to predict just what those uses would be, but I think you could have predicted, even then, that uses would arise. One of the strangest features of the relationship between mathematics and the "real world," but also one of the strongest, is that good mathematics, *whatever its source*, eventually turns out to be useful. There are all sorts of theories why this should be so, ranging from the structure of the human mind to the idea that the universe is somehow built from little bits of mathematics. My feeling is that the answer is probably quite simple: mathematics is the science of patterns, and nature exploits just about every pattern that there is. I admit that I find it much harder to offer a convincing reason for nature to behave in this manner. Maybe the question is back to front: maybe the point is that creatures able to ask that kind of question can evolve only in a universe with that kind of structure.*

Whatever the reasons, mathematics definitely is a useful way to think about nature. What do we want it to tell us about the patterns we observe? There are many answers. We want to understand how they happen; to understand *why* they happen, which is different; to organize the underlying patterns and regularities in the most satisfying way; to predict how nature will behave; to control nature for our own ends; and to make practical use of what we have learned about our world. Mathematics help us to do all these things, and often it is indispensable.

For example, consider the spiral form of a snail shell. *How* the snail makes its shell is largely a matter of chemistry and genetics. Without going into fine points, the snail's genes include recipes for making particular chemicals and instructions for where they should go. Here mathematics lets us do the molecular bookkeeping that makes sense of the different chemical reactions that go on; it describes the atomic structure of the molecules used in shells, it describes the rigidity of shell material as compared to the weakness and pliability of the snail's body, and so on. Indeed, without mathematics we would never

* This explanation, and others, are discussed in *The Collapse of Chaos*, by Jack Cohen and Ian Stewart (New York: Viking, 1994).

have convinced ourselves that matter really is made from atoms, or have worked out how the atoms are arranged. The discovery of genes—and later of the molecular structure of DNA, the genetic material—relied heavily on the existence of mathematical clues. The monk Gregor Mendel noticed tidy numerical relationships in how the proportions of plants with different characters, such as seed color, changed when the plants were crossbred. This led to the basic idea of genetics—that within every organism is some cryptic combination of factors that determines many features of its body plan, and that these factors are somehow shuffled and recombined when passing from parents to offspring. Many different pieces of mathematics were involved in the discovery that DNA has the celebrated double-helical structure. They were as simple as Chargaff's rules: the observation by the Austrian-born biochemist Erwin Chargaff that the four bases of the DNA molecule occur in related proportions; and they are as subtle as the laws of diffraction, which were used to deduce molecular structure from X-ray pictures of DNA crystals.

The question of *why* snails have spiral shells has a very different character. It can be asked in several contexts—in the short-term context of biological development, say, or the long-term context of evolution. The main mathematical feature of the developmental story is the general shape of the spiral. Basically, the developmental story is about the geometry of a creature that behaves in much the same way all the time, but keeps getting bigger. Imagine a tiny animal, with a tiny protoshell attached to it. Then the animal starts to grow. It can grow most easily in the direction along which the open rim of the shell points, because the shell gets in the way if it tries to grow in any other direction. But, having grown a bit, it needs to extend its shell as well, for self-protection. So, of course, the shell grows an extra ring of material around its rim. As this process continues, the animal is getting bigger, so the size of the rim grows. The simplest result is a conical shell, such as you find on a limpet. But if the whole system starts with a bit of a twist, as is quite likely, then the growing edge of the shell rotates slowly as well as expanding, and it rotates in an off-centerd manner. The result is a cone that twists in an ever-expanding spiral. We can use mathematics to relate the resulting geometry to all the different variables—such as growth rate and eccentricity of growth—that are involved.

If, instead, we seek an evolutionary explanation, then we might focus more on the strength of the shell, which conveys an evolutionary advantage, and try to calculate whether a long thin cone is stronger or weaker than a tightly coiled spiral. Or we might be more ambitious and develop mathematical models of the evolutionary process itself, with its combination of random genetic change—that is, mutations—and natural selection.

A remarkable example of this kind of thinking is a computer simulation of the evolution of the eye by Daniel

Nilsson and Susanne Pelger, published in 1994. Recall that conventional evolutionary theory sees changes in animal form as being the result of random mutations followed by subsequent selection of those individuals most able to survive and reproduce their kind. When Charles Darwin announced this theory, one of the first objections raised was that complex structures (like an eye) have to evolve fully formed or else they won't work properly (half an eye is no use at all), but the chance that random mutation will produce a coherent set of complex changes is negligible. Evolutionary theorists quickly responded that while half an eye may not be much use, a half *developed* eye might well be. One with a retina but no lens, say, will still collect light and thereby detect movement; and any way to improve the detection of predators offers an evolutionary advantage to any creature that possesses it. What we have here is a verbal objection to the theory countered by a verbal argument. But the recent computer analysis goes much further.

It starts with a mathematical model of a flat region of cells, and permits various types of "mutation." Some cells may become more sensitive to light, for example, and the shape of the region of cells may bend. The mathematical model is set up as a computer program that makes tiny random changes of this kind, calculates how good the resulting structure is at detecting light and resolving the patterns that it "sees," and selects any changes that improve these abilities. During a simulation that corresponds to a period of about four hundred thousand years—the blink of an eye, in evolutionary terms—the region of cells folds itself up into a deep, spherical cavity with a tiny irislike opening and, most dramatically, a lens. Moreover, like the lenses in our own eyes, it is a lens whose refractive index—the amount by which it bends light—varies from place to place. In fact, the pattern of variation of refractive index that is produced in the computer simulation is very like our own. So here mathematics shows that eyes definitely can evolve gradually and naturally, offering increased survival value at every stage. More than that: Nilsson and Pelger's work demonstrates that given certain key biological faculties (such as cellular receptivity to light, and cellular mobility), structures remarkably similar to eyes *will* form—all in line with Darwin's principle of natural selection. The mathematical model provides a lot of extra detail that the verbal Darwinian argument can only guess at, and gives us far greater confidence that the line of argument is correct.

I said that another function of mathematics is to organize the underlying patterns and regularities in the most satisfying way. To illustrate this aspect, let me return to the question raised in the first chapter. Which—if either—is significant: the three-in-a-row pattern of stars in Orion's belt, or the three-in-a-row pattern to the periods of revolution of Jupiter's satellites? Orion first. Ancient human civilizations organized the stars in the sky in terms of pictures of animals and mythic heroes. In these terms, the alignment of the three stars in Orion appears significant, for otherwise the hero would have no belt from which to hang his sword. However, if we use three-dimensional geometry as an organizing principle and place the three stars in their correct positions in the heavens, then we find that they are at very different distances from the Earth. Their equispaced alignment is an accident, depending on the position from which they are being viewed. Indeed, the very word "constellation" is a misnomer for an arbitrary accident of viewpoint.

The numerical relation between the periods of revolution of Io, Europa, and Ganymede could also be an accident of viewpoint. How can we be sure that "period of revolution" has any significant meaning for nature? However, that numerical relation fits into a dynamical framework in a very significant manner indeed. It is an example of a *resonance*, which is a relationship between periodically moving bodies in which their cycles are locked together, so that they take up the same relative positions at regular intervals. This common cycle time is called the period of the system. The individual bodies may have different—but related—periods. We can work out what this relationship is. When a resonance occurs, all of the participating bodies must return to a standard reference position after a whole number of cycles—but that number can be different for

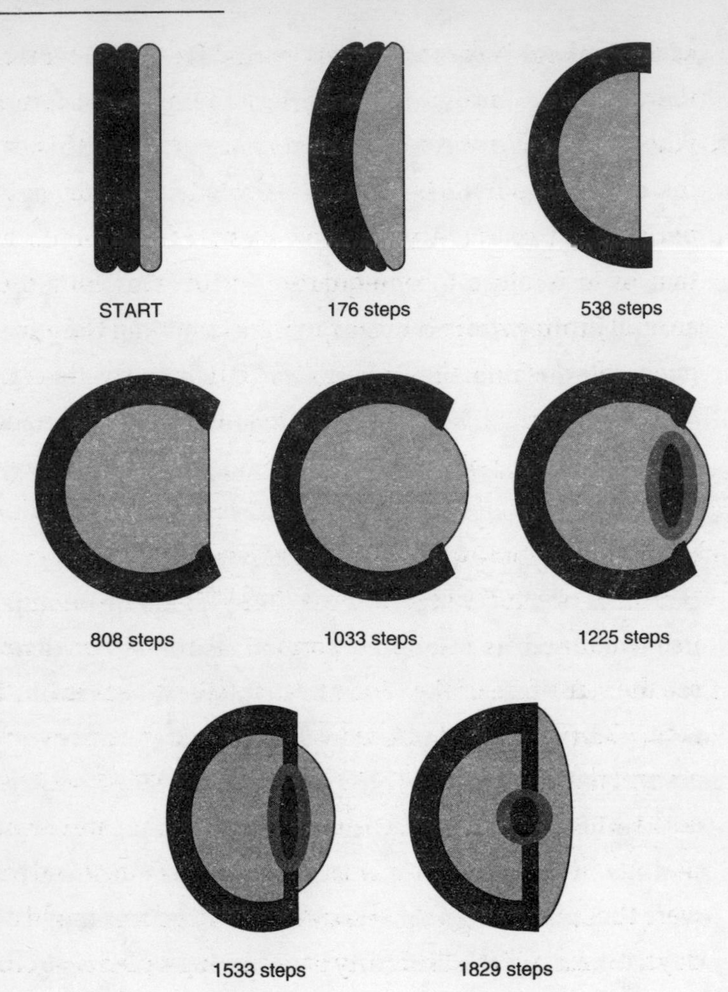

FIGURE 1.
Computer model of the evolution of an eye. Each step in the computation corresponds to about two hundred years of biological evolution.

each. So there is some common period for the system, and therefore each individual body has a period that is some whole-number divisor of the common period. In this case, the common period is that of Ganymede, 7.16 days. The period of Europa is very close to half that of Ganymede, and that of Io is close to one-quarter. Io revolves four times around Jupiter while Europa revolves twice and Ganymede once, after which they are all back in exactly the same relative positions as before. This is called a 4:2:1 resonance.

The dynamics of the Solar System is full of resonances. The Moon's rotational period is (subject to small wobbles caused by perturbations from other bodies) the same as its period of revolution around the Earth—a 1:1 resonance of its orbital and its rotational period. Therefore, we always see the same face of the Moon from the Earth, never its "far side." Mercury rotates once every 58.65 days and revolves around the Sun every 87.97 days. Now, $2 \times 87.97 = 175.94$, and $3 \times 58.65 = 175.95$, so Mercury's rotational and orbital periods are in a 2:3 resonance. (In fact, for a long time they were thought to be in 1:1 resonance, both being roughly 88 days, because of the difficulty of observing a planet as close to the Sun as Mercury is. This gave rise to the belief that one side of Mercury is incredibly hot and the other incredibly cold, which turns out not to be true. A resonance, however, there is—and a more interesting one than mere equality.)

In between Mars and Jupiter is the asteroid belt, a broad zone containing thousands of tiny bodies. They are not uniformly distributed. At certain distances from the Sun we find asteroid "beltlets"; at other distances we find hardly any. The explanation—in both cases—is resonance with Jupiter. The Hilda group of asteroids, one of the beltlets, is in 2:3 resonance with Jupiter. That is, it is at just the right distance so that all of the Hilda asteroids circle the Sun three times for every two revolutions of Jupiter. The most noticeable gaps are at 2:1, 3:1, 4:1, 5:2, and 7:2 resonances. You may be worried that resonances are being used to explain both clumps and gaps. The reason is that each resonance has its own idiosyncratic dynamics; some cause clustering, others do the opposite. It all depends on the precise numbers.

Another function of mathematics is prediction. By understanding the motion of heavenly bodies, astronomers could predict lunar and solar eclipses and the return of comets. They knew where to point their telescopes to find asteroids that had passed behind the Sun, out of observational contact. Because the tides are controlled mainly by the position of the Sun and Moon relative to the Earth, they could predict tides many years ahead. (The chief complicating factor in making such predictions is not astronomy: it is the shape of the continents and the profile of the ocean depths, which can delay or advance a high tide. However, these stay pretty much the same from one century to the next, so that once their effects have been understood it is a routine task to compensate for them.) In contrast, it is much harder to predict the weather. We know just as much about the mathematics of weather as we do about the mathematics of tides, but weather has an inherent unpredictability. Despite this, meteorologists can make effective short-term predictions of weather patterns—say, three or four days in advance. The unpredictability of the weather, however, has nothing at all to do with randomness—a topic we will take up in chapter 8, when we discuss the concept of chaos.

The role of mathematics goes beyond mere prediction. Once you understand how a system works, you don't have to remain a passive observer. You can attempt to control the system, to make it do what you want. It pays not to be too ambitious: weather control, for example, is in its infancy—we can't make rain with any great success, even when there are rainclouds about. Examples of control systems range from the thermostat on a boiler, which keeps it at a fixed temperature, to the medieval practice of coppicing woodland. Without a sophisticated mathematical control system, the space shuttle would fly like the brick it is, for no human pilot can respond quickly enough to correct its inherent instabilities. The use of electronic pacemakers to help people with heart disease is another example of control.

These examples bring us to the most down-to-earth aspect of mathematics: its practical applications—how mathematics earns its keep. Our world rests on mathematical foundations, and mathematics is unavoidably embedded in our global culture. The only reason we don't always realize just how strongly our lives are affected by mathematics is that, for sensible reasons, it is kept as far as possible behind the scenes. When you go to the travel agent and book a vacation, you don't need to understand the intricate mathematical and physical theories that make it possible to design computers and telephone lines, the optimization routines that schedule as many flights as possible around any particular airport, or the signal-processing methods used to provide accurate radar images for the pilots. When you watch a television program, you don't need to understand the three-dimensional geometry used to produce special effects on the screen, the coding methods used to transmit TV signals by satellite, the mathematical methods used to solve the equation for the orbital motion of the satellite, the thousands of different applications of mathematics during every step of the manufacture of every component of the spacecraft that launched the satellite into position. When a farmer plants a new strain of potatoes, he does not need to know the statistical theories of genetics that identified which genes made that particular type of plant resistant to disease.

But somebody had to understand all these things in the past, otherwise airliners, television, spacecraft, and disease-resistant potatoes wouldn't have been invented. And somebody has to understand all these things now, too, otherwise they won't continue to function. And somebody has to be inventing new mathematics in the future, able to solve problems that either have not arisen before or have hitherto proved intractable, otherwise our society will fall apart when change requires solutions to new problems or new solutions to old problems. If mathematics, including everything that rests on it, were somehow suddenly to be withdrawn from our world, human society would collapse in an instant. And if mathematics were to be frozen, so that it never went a single step farther, our civilization would start to go backward.

We should not expect new mathematics to give an immediate dollars-and-cents payoff. The transfer of a mathematical idea into something that can be made in a factory or used in a home generally takes time. Lots of time: a century is not unusual. In chapter 5, we will see how seventeenth-century interest in the vibrations of a violin string led, three hundred years later, to the discovery of radio waves and the invention of radio, radar, and television. It might have been done quicker, but not *that* much quicker. If you think—as many people in our increasingly managerial culture do—that the process of scientific discovery can be speeded up by focusing on the application as a goal and ignoring "curiosity-driven" research, then you are wrong. In fact that very phrase, "curiosity-driven research," was introduced fairly recently by unimaginative bureaucrats as a deliberate put-down. Their desire for tidy projects offering guaranteed short-term profit is much too simple-minded, because goal-oriented research can deliver only predictable results. You have to be able to see the goal in order to aim at it. But anything you can see, your competitors can see, too. The pursuance of safe research will impoverish us all. The really important breakthroughs are always unpredictable. It is their very unpredictability that makes them important: they change our world in ways we didn't see coming.

Moreover, goal-oriented research often runs up against a brick wall, and not only in mathematics. For example, it took approximately eighty years of intense engineering effort to develop the photocopying machine after the basic principle of xerography had been discovered by scientists. The first fax machine was invented over a century ago, but it didn't work fast enough or reliably enough. The principle of holography (three-dimensional pictures, see your credit card) was discovered over a century ago, but nobody then knew how to produce the necessary beam of coherent light—light with all its waves in step. This kind of delay is not at all unusual in industry, let alone in more intellectual areas of research, and the impasse is usually broken only when an unexpected new idea arrives on the scene.

There is nothing wrong with goal-oriented research as a way of achieving specific feasible goals. But the dreamers and the mavericks must be allowed some free rein, too. Our world is not static: new problems constantly arise, and old answers often stop working. Like Lewis Carroll's Red Queen, we must run very fast in order to stand still.

CHAPTER 3

WHAT MATHEMATICS IS ABOUT

When we hear the word "mathematics," the first thing that springs to mind is numbers. Numbers are the heart of mathematics, an all-pervading influence, the raw materials out of which a great deal of mathematics is forged. But numbers on their own form only a tiny part of mathematics. I said earlier that we live in an intensely mathematical world, but that whenever possible the mathematics is sensibly tucked under the rug to make our world "user-friendly." However, some mathematical ideas are so basic to our world that they cannot stay hidden, and numbers are an especially prominent example. Without the ability to count eggs and subtract change, for instance, we could not even buy food. And so we teach arithmetic. To everybody. Like reading and writing, its absence is a major handicap. And that creates the overwhelming impression that mathematics is mostly a matter of numbers—which isn't really true. The numerical tricks we learn in arithmetic are only the tip of an iceberg. We can run our everyday lives without much more, but our culture cannot run our society by using such limited ingredients. Numbers are just one type of object that mathematicians think about. In this chapter, I will try to show you some of the others and explain why they, too, are important.

Inevitably my starting point has to be numbers. A large part of the early prehistory of mathematics can be summed up as the discovery, by various civilizations, of a wider and wider range of things that deserved to be called numbers. The simplest are the numbers we use for counting. In fact, counting began long before there were symbols like 1,2,3, because it is possible to count without using numbers at all—say, by counting on your fingers. You can work out that "I have two hands and a thumb of camels" by folding down fingers as your eye glances over the camels. You don't actually have to have the concept of the number "eleven" to keep track of whether anybody is stealing your camels. You just have to notice that next time you seem to have only two hands of camels—so a thumb of camels is missing.

You can also record the count as scratches on pieces of wood or bone. Or you can make tokens to use as counters—clay disks with pictures of sheep on them for counting sheep, or disks with pictures of camels on them for counting camels. As the animals parade past you, you drop tokens into a bag—one token for each animal. The use of symbols for numbers probably developed about five thousand years ago, when such counters were wrapped in a clay envelope. It was a nuisance to break open the clay covering every time the accountants wanted to check the contents, and to make another one when they had finished. So people put special marks on the outside of the envelope summarizing what was inside. Then they realized that they didn't actually need any counters inside at all: they could just make the same marks on clay tablets.

It's amazing how long it can take to see the obvious. But of course it's only obvious *now.*

The next invention beyond counting numbers was the kind of number we now symbolize as 2/3 (two thirds) or 22/7 (twenty-two sevenths—or, equivalently, three and one-seventh). You can't count with fractions—although two-thirds of a camel might be edible, it's not countable—but you can do much more interesting things instead. In particularly, if three brothers inherit two camels between them, you can think of each as owning two-thirds of a camel—a convenient legal fiction, one with which we are so comfortable that we forget how curious it is if taken literally.

Much later, between 400 and 1200 AD, the concept of zero was invented and accepted as denoting a number. If you think that the late acceptance of zero as a number is strange, bear in mind that for a long time "one" was not considered a number because it was thought that a number of things ought to be several of them. Many history books say that the key idea here was the invention of a symbol for "nothing." That may have been the key to making arithmetic practical; but for mathematics the important idea was the concept of a new kind of number, one that *represented* the concrete idea "nothing." Mathematics uses symbols, but it no more *is* those symbols than music is musical notation or language is strings of letters from an alphabet. Carl Friedrich Gauss, thought by many to be the greatest mathematician ever to have lived, once said (in Latin) that what matters in mathematics is "not notations, but notions." The pun "non notationes, sed notiones" worked in Latin, too.

The next extension of the number concept was the invention of negative numbers. Again, it makes little sense to think of minus two camels in a literal sense; but if you owe somebody two camels, the number you own is effectively diminished by two. So a negative number can be thought of as representing a debt. There are many different ways to interpret these more esoteric kinds of number; for instance, a negative temperature (in degrees Celsius) is one that is colder than freezing, and an object with negative velocity is one that is moving backward. So the same abstract mathematical object may represent more than one aspect of nature.

Fractions are all you need for most commercial transactions, but they're not enough for mathematics. For example, as the ancient Greeks discovered to their chagrin, the square root of two is not exactly representable as a fraction. That is, if you multiply any fraction by itself, you won't get two *exactly*. You can get very close—for example, the square of 17/12 is 289/144, and if only it were 288/144 you would get two. But it isn't, and you don't—and whatever fraction you try, you never will. The square root of two, usually denoted $\sqrt{2}$, is therefore said to be "irrational." The simplest way to enlarge the number system to include the irrationals is to use the so-called real numbers—a breathtakingly inappropriate name, inasmuch as they are represented by decimals that go on forever, like 3.14159 . . . , where the dots indicate an infinite number of digits. How can things be real if you can't even write them down fully? But the name stuck, probably because real numbers formalize many of our natural visual intuitions about lengths and distances.

The real numbers are one of the most audacious idealizations made by the human mind, but they were used happily for centuries before anybody worried about the logic behind them. Paradoxically, people worried a great deal about the next enlargement of the number system, even though it was entirely harmless. That was the introduction of square roots for negative numbers, and it led to the "imaginary" and "complex" numbers. A professional mathematican should never leave home without them, but fortunately nothing in this book will require knowledge of complex numbers, so I'm going to tuck them under the mathematical carpet and hope you don't notice. However, I should point out that it is easy to interpret an infinite decimal as a sequence of ever-finer approximations to some measurement—say, of a length or a weight—whereas a comfortable interpretation of the square root of minus one is more elusive.

In current terminology, the whole numbers 0,1,2,3, . . . are known as the natural numbers. If negative whole numbers are included, we have the integers. Positive and negative fractions are called rational numbers. Real numbers are more general; complex numbers more general still. So here we have five systems, each more inclusive than the previous: natural numbers, integers, rationals, real numbers, and complex numbers. In this book, the important number systems will be the integers and the reals. We'll need to talk about rational numbers every so often; and as I've just said, we can ignore the complex numbers altogether. But I hope you understand by now that the word "number" does not have any immutable god-given meaning. More than once the scope of that word was extended, a process that in principle might occur again at any time.

However, mathematics is not just about numbers. We've already had a passing encounter with a different kind of object of mathematical thought, an *operation*; examples are addition, subtraction, multiplication, and division. In general, an operation is something you apply to two (sometimes more) mathematical objects to get a third object. I also alluded to a third type of mathematical object when I mentioned square roots. If you start with a number and form its square root, you get another number. The term for such an "object" is *function*. You can think of a function as a mathematical rule that starts with a mathematical object—usually a number—and associates to it another object in a specific manner. Functions are often defined using algebraic formulas, which are just shorthand ways to explain what the rule is, but they can be defined by any convenient method. Another term with the same meaning as "function" is *transformation*: the rule transforms the first object into the second. This term tends to be used when the rules are geometric, and in chapter 6 we will use transformations to capture the mathematical essence of symmetry.

Operations and functions are very similar concepts. Indeed, on a suitable level of generality there is not much to distinguish them. Both of them are processes rather than things. And now is a good moment to open up Pandora's box and explain one of the most powerful general weapons in the mathematician's armory, which we might call the "thingification of processes." (There is a dictionary term, *reification*, but it sounds pretentious.) Mathematical "things" have no existence in the real world: they are abstractions. But mathematical processes are also abstractions, so processes are no less "things" than the "things" to which they are applied. The thingification of processes is commonplace. In fact, I can make out a very good case that the number "two" is not actually a thing but a process—the process you carry out when you associate two camels or two sheep with the symbols "1,2" chanted in turn. A number is a process that has long ago been thingified so thoroughly that everybody thinks of it as a thing. It is just as feasible—though less familiar to most of us—to think of an operation or a function as a thing. For example, we might talk of "square root" as if it were a thing—and I mean here not the square root of any particular number, but the function itself. In this image, the square-root function is a kind of sausage machine: you stuff a number in at one end and its square root pops out at the other.

In chapter 6, we will treat motions of the plane or space as if they are things. I'm warning you now because you may find it disturbing when it happens. However, mathematicians aren't the only people who play the thingification game. The legal profession talks of "theft" as if it were a thing; it even knows what kind of thing it is—a crime. In phrases such as "two major evils in Western society are drugs and theft" we find one genuine thing and one thingified thing, both treated as if they were on exactly the same level. For theft is a process, one whereby my property is transferred without my agreement to somebody else, but drugs have a real physical existence.

Computer scientists have a useful term for things that can be built up from numbers by thingifying processes: they call them data structures. Common examples in computer science are lists (sets of numbers written in sequence) and arrays (tables of numbers with several rows and columns).

I've already said that a picture on a computer screen can be represented as a list of pairs of numbers; that's a more complicated but entirely sensible data structure. You can imagine much more complicated possibilities—arrays that are tables of lists, not tables of numbers; lists of arrays; arrays of arrays; lists of lists of arrays of lists . . . Mathematics builds its basic objects of thought in a similiar manner. Back in the days when the logical foundations of mathematics were still being sorted out, Bertrand Russell and Alfred North Whitehead wrote an enormous three-volume work, *Principia Mathematica*, which began with the simplest possible logical ingredient—the idea of a set, a collection of things. They then showed how to build up the rest of mathematics. Their main objective was to analyze the logical structure of mathematics, but a major part of their effort went into devising appropriate data structures for the important objects of mathematical thought.

The image of mathematics raised by this description of its basic objects is something like a tree, rooted in numbers and branching into ever more esoteric data structures as you proceed from trunk to bough, bough to limb, limb to twig . . . But this image lacks an essential ingredient. It fails to describe how mathematical concepts interact. Mathematics is not just a collection of isolated facts: it is more like a landscape; it has an inherent geography that its users and creators employ to navigate through what would otherwise be an impenetrable jungle. For instance, there is a metaphorical feeling of distance. Near any particular mathematical fact we find other, related facts. For example, the fact that the circumference of a circle is π (pi) times its diameter is very close to the fact that the circumference of a circle is 2π times its radius. The connection between these two facts is immediate: the diameter is twice the radius. In contrast, unrelated ideas are more distant from each other; for example, the fact that there are exactly six different ways to arrange three objects in order is a long way away from facts about circles. There is also a metaphorical feeling of prominence. Soaring peaks pierce the sky—important ideas that can be used widely and seen from far away, such as Pythagoras's theorem about right triangles, or the basic techniques of calculus. At every turn, new vistas arise—an unexpected river that must be crossed using stepping stones, a vast, tranquil lake, an impassable crevasse. The user of mathematics walks only the well-trod parts of this mathematical territory. The creator of mathematics explores its unknown mysteries, maps them, and builds roads through them to make them more easily accessible to everybody else.

The ingredient that knits this landscape together is *proof*. Proof determines the route from one fact to another. To professional mathematicians, no statement is considered valid unless it is proved beyond any possibility of logical error. But there are limits to what can be proved, and how it can be proved. A great deal of work in philosophy and the foundations of mathematics has established that you can't prove everything, because you have to start somewhere; and even when you've decided where to start, some statements may be neither provable nor disprovable. I don't want to explore those issues here; instead, I want to take a pragmatic look at what proofs are and why they are needed.

Textbooks of mathematical logic say that a proof is a sequence of statements, each of which either follows from previous statements in the sequence or from agreed axioms—unproved but explicitly stated assumptions that in effect define the area of mathematics being studied. This is about as informative as describing a novel as a sequence of sentences, each of which either sets up an agreed context or follows credibly from previous sentences. Both definitions miss the essential point: that both a proof and a novel must tell an interesting story. They do capture a secondary point, that the story must be convincing, and they also describe the overall format to be used, but a good story line is the most important feature of all.

Very few textbooks say that.

Most of us are irritated by a movie riddled with holes, however polished its technical production may be. I saw one recently in which an airport is taken over by guerrillas who shut down the electronic equipment used by the control tower and substitute their own. The airport authorities and the hero then spend half an hour or more of movie time—several hours of story time—agonizing about their inability to communicate with approaching aircraft, which are stacking up in the sky overhead and running out of fuel. It occurs to no one that there is a second, fully functioning airport no more than thirty miles away, nor do they think to telephone the nearest Air Force base. The story was brilliantly and expensively filmed—and silly.

That didn't stop a lot of people from enjoying it: their critical standards must have been lower than mine. But we all have limits to what we are prepared to accept as credible. If in an otherwise realistic film a child saved the day by picking up a house and carrying it away, most of us would lose interest. Similarly, a mathematical proof is a story about mathematics that works. It does not have to dot every *i* and cross every *t*; readers are expected to fill in routine steps for themselves—just as movie characters may suddenly appear in new surroundings without it being necessary to show how they got there. But the story must not have gaps, and it certainly must not have an unbelievable plot line. The rules are stringent: in mathematics, a single flaw is fatal. Moreover, a subtle flaw can be just as fatal as an obvious one.

Let's take a look at an example. I have chosen a simple one, to avoid technical background; in consequence, the proof tells a simple and not very significant story. I stole it from a colleague, who calls it the SHIP/DOCK Theorem. You probably know the type of puzzle in which you are given one word (SHIP) and asked to turn it into another word (DOCK) by changing one letter at a time and getting a valid word at every stage. You might like to try to solve this one before reading on: if you do, you will probably understand the theorem, and its proof, more easily.

Here's one solution:

SHIP
SLIP
SLOP
SLOT
SOOT
LOOT
LOOK
LOCK
DOCK

There are plenty of alternatives, and some involve fewer words. But if you play around with this problem, you will eventually notice that all solutions have one thing in common; at least one of the intermediate words must contain two vowels.

O.K., so prove it.

I'm not willing to accept experimental evidence. I don't care if you have a hundred solutions and every single one of them includes a word with two vowels. You won't be happy with such evidence, either, because you will have a sneaky feeling that you may just have missed some really clever sequence that doesn't include such a word. On the other hand, you will probably also have a distinct feeling that somehow "it's obvious." I agree: but *why* is it obvious?

You have now entered a phase of existence in which most mathematicians spend most of their time: frustration. You know what you want to prove, you believe it, but you don't see a convincing story line for a proof. What this means is that you are lacking some key idea that will blow the whole problem wide open. In a moment I'll give you a hint. Think about it for a few minutes, and you will probably experience a much more satisfying phase of the mathematician's existence: illumination.

Here's the hint. Every valid word in English must contain a vowel.

It's a very simple hint. First, convince yourself that it's true. (A dictionary search is acceptable, provided it's a big dictionary.) Then consider its implications . . .

O.K., either you got it or you've given up. Whichever of these you did, all professional mathematicians have done the same on a lot of their problems. Here's the trick. You have to concentrate on what happens to the vowels. Vowels are the peaks in the SHIP/DOCK landscape, the landmarks between which the paths of proof wind.

In the initial word SHIP there is only one vowel, in the third position. In the final word DOCK there is also only one vowel, but in the second position. How does the vowel change position? There are three possibilities. It may hop from one location to the other; it may disappear altogether and reappear later on; or an extra vowel or vowels may be created and subsequently eliminated.

The third possibility leads pretty directly to the theorem. Since only one letter at a time changes, at some stage the word must change from having one vowel to having two. It can't leap from having one vowel to having three, for example. But what about the other possibilities? The hint that I mentioned earlier tells us that the single vowel in SHIP cannot disappear altogether. That leaves only the first possibility: that there is always one vowel, but it hops from position 3 to position 2. However, that can't be done by changing only one letter! You have to move, in one step, from a vowel at position 3 and a consonant at position 2 to a consonant at position 3 and a vowel at position 2. That implies that two letters must change, which is illegal. Q.E.D., as Euclid used to say.

A mathematician would write the proof out in a much more formal style, something like the textbook model, but the important thing is to tell a convincing story. Like any good story, it has a beginning and an end, and a story line that gets you from one to the other without any logical holes appearing. Even though this is a very simple example, and it isn't standard mathematics at all, it illustrates the essentials: in particular, the dramatic difference between an argument that is genuinely convincing and a hand-waving argument that sounds plausible but doesn't really gel. I hope it also put you through some of the emotional experiences of the creative mathematician: frustration at the intractability of what ought to be an easy question, elation when light dawned, suspicion as you checked whether there were any holes in the argument, aesthetic satisfaction when you decided the idea really was O.K. and realized how neatly it cut through all the apparent complications. Creative mathematics is just like this—but with more serious subject matter.

Proofs must be convincing to be accepted by mathematicians. There have been many cases where extensive numerical evidence suggested a completely wrong answer. One notorious example concerns prime numbers—numbers that have no divisors except themselves and 1. The sequence of primes begins 2,3,5,7,11,13,17,19 and goes on forever. Apart from 2, all primes are odd; and the odd primes fall into two classes: those that are one less than a multiple of four (such as 3,7,11,19) and those that are one more than a multiple of four (such as 5,13,17). If you run along the sequence of primes and count how many of them fall into each class, you will observe that there always seem to be more primes in the "one less" class than in the "one more" class. For example, in the list of the seven pertinent primes above, there are four primes in the first class but only three in the second. This pattern persists for numbers up to at least a trillion, and it seems entirely reasonable to conjecture that it is always true.

However, it isn't.

By indirect methods, number theorists have shown that when the primes get sufficiently big, the pattern changes and the "one more than a multiple of four" class goes into the lead. The first proof of this fact worked only when the numbers got bigger than 10'10'10'10'46, where to avoid giving the printer kittens I've used the ' sign to indicate forming a power. This number is utterly gigantic. Written out in full, it would go 10000. . . 000, with a very large number of 0s. If all the matter in the universe were turned into paper, and a zero could be inscribed on every electron, there wouldn't be enough of them to hold even a tiny fraction of the necessary zeros.

No amount of experimental evidence can account for the possibility of exceptions so rare that you need numbers that big to locate them. Unfortunately, even rare exceptions matter in mathematics. In ordinary life, we seldom worry about things that might occur on one occasion out of a trillion. Do you worry about being hit by a meteorite? The odds are about one in a trillion. But mathematics piles logical deductions on top of each other, and if any step is wrong the whole edifice may tumble. If you have stated as a fact that all numbers behave in some manner, and there is just one that does not, then you are wrong, and everything you have built on the basis of that incorrect fact is thrown into doubt.

Even the very best mathematicians have on occasion claimed to have proved something that later turned out not to be so—their proof had a subtle gap, or there was a simple error in a calculation, or they inadvertently assumed something that was not as rock-solid as they had imagined. So, over the centuries, mathematicians have learned to be extremely critical of proofs. Proofs knit the fabric of mathematics together, and if a single thread is weak, the entire fabric may unravel.

CHAPTER 4

THE CONSTANTS OF CHANGE

For a good many centuries, human thought about nature has swung between two opposing points of view. According to one view, the universe obeys fixed, immutable laws, and everything exists in a well-defined objective reality. The opposing view is that there is no such thing as objective reality; that all is flux, all is change. As the Greek philosopher Heraclitus put it, "You can't step into the same river twice." The rise of science has largely been governed by the first viewpoint. But there are increasing signs that the prevailing cultural background is starting to switch to the second—ways of thinking as diverse as postmodernism, cyberpunk, and chaos theory all blur the alleged objectiveness of reality and reopen the ageless debate about rigid laws and flexible change.

What we really need to do is get out of this futile game altogether. We need to find a way to step back from these opposing worldviews—not so much to seek a synthesis as to see them both as two shadows of some higher order of reality— shadows that are different only because the higher order is being seen from two different directions. But does such a higher order exist, and if so, is it accessible? To many—especially scientists—Isaac Newton represents the triumph of rationality over myticism. The famous economist John Maynard Keynes, in his essay *Newton, the Man*, saw things differently:

> In the eighteenth century and since, Newton came to be thought of as the first and greatest of the modern age of scientists, a rationalist, one who taught us to think on the lines of cold and untinctured reason. I do not see him in this light. I do not think that anyone who has pored over the contents of that box which he packed up when he finally left Cambridge in 1696 and which, though partly dispersed, have come down to us, can see him like that. Newton was not the first of the age of reason. He was the last of the magicians, the last of the Babylonians and Sumerians, the last great mind which looked out on the visible and intellectual world with the same eyes as those who began to build our intellectual inheritance rather less than 10,000 years ago. Isaac Newton, a posthumous child born with no father on Christmas Day, 1642, was the last wonder-child to whom the Magi could do sincere and appropriate homage.

Keynes was thinking of Newton's personality, and of his interests in alchemy and religion as well as in mathematics and physics. But in Newton's mathematics we also find the first significant step toward a worldview that transcends and unites both rigid law and flexible flux. The universe may appear to be a storm-tossed ocean of change, but

Newton—and before him Galileo and Kepler, the giants upon whose shoulders he stood—realized that change obeys rules. Not only can law and flux coexist, but law *generates* flux.

Today's emerging sciences of chaos and complexity supply the missing converse: flux generates law. But that is another story, reserved for the final chapter.

Prior to Newton, mathematics had offered an essentially static model of nature. There are a few exceptions, the most obvious being Ptolemy's theory of planetary motion, which reproduced the observed changes very accurately using a system of circles revolving about centers that themselves were attached to revolving circles—wheels within wheels within wheels. But at that time the perceived task of mathematics was to discover the catalogue of "ideal forms" employed by nature. The circle was held to be the most perfect shape possible, on the basis of the democratic observation that every point on the circumference of a circle lies at the same distance from its center. Nature, the creation of higher beings, is by definition perfect, and ideal forms are mathematical perfection, so of course the two go together. And perfection was thought to be unblemished by change.

Kepler challenged that view by finding ellipses in place of complex systems of circles. Newton threw it out altogether, replacing forms by the laws that produce them.

Although its ramifications are immense, Newton's

approach to motion is a simple one. It can be illustrated using the motion of a projectile, such as a cannonball fired from a gun at an angle. Galileo discovered experimentally that the path of such a projectile is a parabola, a curve known to the ancient Greeks and related to the ellipse. In this case, it forms an inverted U-shape. The parabolic path can be most easily understood by decomposing the projectile's motion into two independent components: motion in a horizontal direction and motion in a vertical direction. By thinking about these two types of motion separately, and putting them back together only when each has been understood in its own right, we can see why the path should be a parabola.

The cannonball's motion in the horizontal direction, parallel to the ground, is very simple: it takes place at a constant speed. Its motion in the vertical direction is more interesting. It starts moving upward quite rapidly, then it slows down, until for a split second it appears to hang stationary in the air; then it begins to drop, slowly at first but with rapidly increasing velocity.

Newton's insight was that although the position of the cannonball changes in quite a complex way, its velocity changes in a much simpler way, and its acceleration varies in a very simple manner indeed. Figure 2 summarizes the relationship between these three functions, in the following example.

Suppose for the sake of illustration that the initial

upward velocity is fifty meters per second (50 m/sec). Then the height of the cannonball above ground, at one-second intervals, is:

0,45,80,105,120,125,120,105,80,45,0.

You can see from these numbers that the ball goes up, levels off near the top, and then goes down again. But the general pattern is not entirely obvious. The difficulty was compounded in Galileo's time—and, indeed, in Newton's—because it was hard to measure these numbers directly. In actual fact, Galileo rolled a ball up a gentle slope to slow the whole process down. The biggest problem was to measure time accurately: the historian Stillman Drake has suggested that perhaps Galileo hummed tunes to himself and subdivided the basic beat in his head, as a musician does.

The pattern of distances is a puzzle, but the pattern of velocities is much clearer. The ball starts with an upward velocity of 50 m/sec. One second later, the velocity has decreased to (roughly) 40 m/sec; a second after that, it is 30 m/sec; then 20 m/sec, 10 m/sec, then 0 m/sec (stationary). A second after that, the velocity is 10 m/sec *downward*. Using negative numbers, we can think of this as an upward velocity of –10 m/sec. In successive seconds, the pattern continues: –20 m/sec, –30 m/sec, –40 m/sec, –50 m/sec. At this point, the cannonball hits the ground. So the sequence of velocities, measured at one-second intervals, is:

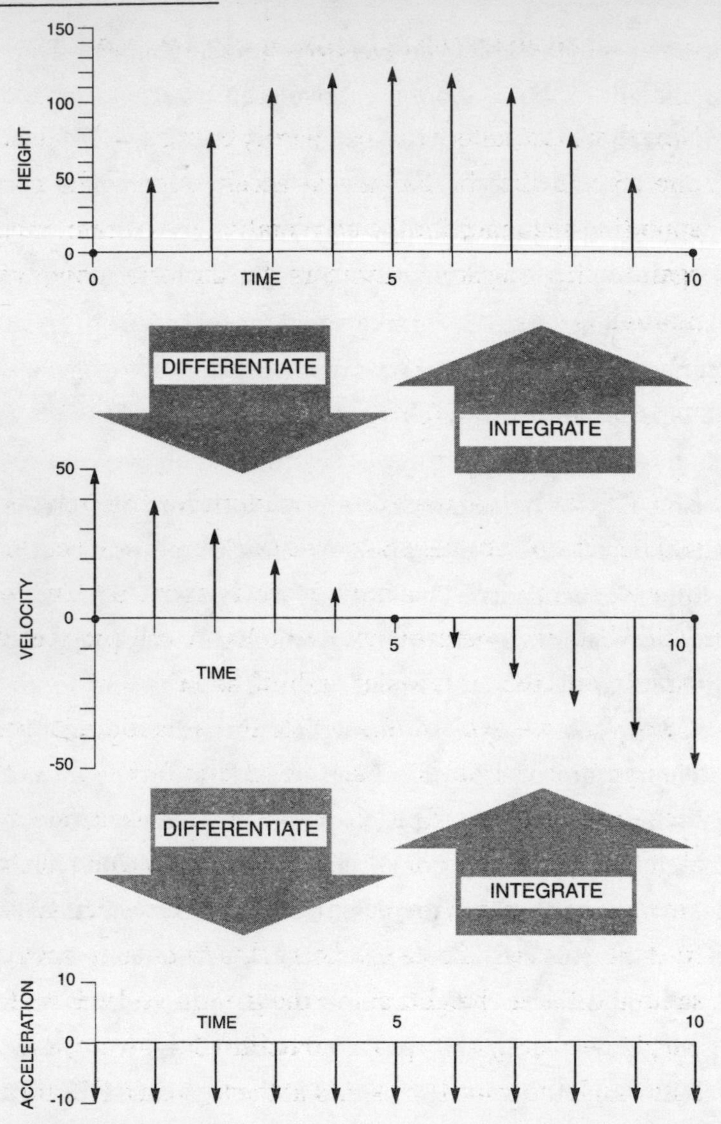

FIGURE 2.
Calculus in a nutshell. Three mathematical patterns determined by a cannonball: height, velocity, and acceleration. The pattern of heights, which is what we naturally observe, is complicated. Newton realized that the pattern of velocities is simpler, while the pattern of accelerations is simpler still. The two basic operations of calculus, differentiation and integration, let us pass from any of these patterns to any other. So we can work with the simplest, acceleration, and deduce the one we really want – height.

50,40,30,20,10,0,–10,–20,–30,–40,–50.

Now there is a pattern that can hardly be missed; but let's go one step further by looking at accelerations. The corresponding sequence for the acceleration of the cannonball, again using negative numbers to indicate downward motion, is

–10,–10,–10,–10,–10,–10,–10,–10,–10,–10,–10.

I think you will agree that the pattern here is extremely simple. The ball undergoes a *constant* downward acceleration of 10 m/sec^2. (The true figure is about 9.81 m/sec^2, depending on whereabouts on the Earth you perform the experiment. But 10 is easier to think about.)

How can we explain this constant that is hiding among the dynamic variables? When all else is flux, why is the acceleration fixed? One attractive explanation has two elements. The first is that the Earth must be pulling the ball downward; that is, there is a gravitational force that acts on the ball. It is reasonable to expect this force to remain the same at different heights above the ground. Indeed, we feel weight because gravity pulls our bodies downward, and we still weigh the same if we stand at the top of a tall building. Of course, this appeal to everyday observation does not tell us what happens if the distance becomes sufficiently large—say the distance that separates the Moon from the

Earth. That's a different story, to which we shall return shortly.

The second element of the explanation is the real breakthrough. We have a body moving under a constant downward force, and we observe that it undergoes a constant downward acceleration. Suppose, for the sake of argument, that the pull of gravity was a lot stronger: then we would expect the downward acceleration to be a lot stronger, too. Without going to a heavy planet, such as Jupiter, we can't test this idea, but it looks reasonable; and it's equally reasonable to suppose that on Jupiter the downward acceleration would again be constant—but a different constant from what it is here. The simplest theory consistent with this mixture of real experiments and thought experiments is that when a force acts on a body, the body experiences an acceleration that is proportional to that force. And this is the essence of Newton's law of motion. The only missing ingredients are the assumption that this is always true, for all bodies and for all forces, whether or not the forces remain constant; and the identification of the constant of proportionality as being related to the mass of the body. To be precise, Newton's law of motion states that

$$\text{mass} \times \text{acceleration} = \text{force}.$$

That's it. Its great virtue is that it is valid for any system of masses and forces, including masses and forces that change over time. We could not have anticipated this universal applicability from the argument that led us to the law; but it turns out to be so.

Newton stated three laws of motion, but the modern approach views them as three aspects of a single mathematical equation. So I will use the phrase "Newton's law of motion" to refer to the whole package.

The mountaineer's natural urge when confronted with a mountain is to climb it; the mathematician's natural urge when confronted with an equation is to solve it. But how? Given a body's mass and the forces acting on it, we can easily solve this equation to get the acceleration. But this is the answer to the wrong question. Knowing that the acceleration of a cannonball is always -10 m/sec^2 doesn't tell us anything obvious about the shape of its trajectory. This is when the branch of mathematics known as calculus comes in; indeed it is why Newton (and Leibniz) invented it. Calculus provides a technique, which nowadays is called integration, that allows us to move from knowledge of acceleration at any instant to knowledge of velocity at any instant. By repeating the same trick, we can then obtain knowledge of position at any instant. And that is the answer to the *right* question.

As I said earlier, velocity is rate of change of position, and acceleration is rate of change of velocity. Calculus is a mathematical scheme invented to handle questions about rates of change. In particular, it provides a technique for *finding* rates of change—a technique known as differentiation. Integration "undoes" the effect of differentiation; and integrating twice undoes the effect of differentiating twice. Like the twin faces of the Roman god Janus, these twin techniques of calculus point in opposite directions. Between them, they tell you that if you know any one of the functions—position, velocity, or acceleration—at every instant, then you can work out the other two.

Newton's law of motion teaches an important lesson: namely, that the route from nature's laws to nature's behavior need not be direct and obvious. Between the behavior we observe and the laws that produce it is a crevasse, which the human mind can bridge only by mathematical calculations. This is not to suggest that nature *is* mathematics—that (as the physicist Paul Dirac put it) "God is a mathematician." Maybe nature's patterns and regularities have other origins; but, at the very least, mathematics is an extremely effective way for human beings to come to grips with those patterns.

All of the laws of physics that were discovered by pursuing Isaac Newton's basic insight—that change in nature can be described by mathematical processes, just as form in nature can be described by mathematical things—have a similar character. The laws are formulated as equations that relate not the physical quantities of primary interest but the rates at which those quantities change with time, or the rates at which those rates change with time. For example the "heat equation," which determines how heat flows through a conducting body, is all about the rate of change of the body's temperature; and the "wave equation," which governs the motion of waves in water, air, or other materials, is about the rate of change of the rate of change of the height of the wave. The physical laws for light, sound, electricity, magnetism, the elastic bending of materials, the flow of fluids, and the course of a chemical reaction, are all equations for various rates of change.

Because a rate of change is about the difference between some quantity now and its value an instant into the future, equations of this kind are called *differential equations.* The term "differentiation" has the same origin. Ever since Newton, the strategy of mathematical physics has been to describe the universe in terms of differential equations, and then solve them.

However, as we have pursued this strategy into more sophisticated realms, the meaning of the word "solve" has undergone a series of major changes. Originally it implied finding a precise mathematical formula that would describe what a system does at any instant of time. Newton's discovery of another important natural pattern, the law of gravitation, rested upon a solution of this kind. He began with Kepler's discovery that planets move in ellipses, together with two other mathematical regularities that were also noted by Kepler. Newton asked what kind of force, acting on a planet, would be needed to produce the pattern that Kepler had found. In effect, Newton was trying to work backward from behavior to laws, using a process of induction rather than deduction. And he discovered a very beautiful result. The necessary force should always point in the direction of the Sun; and it should decrease with the distance from the planet to the Sun. Moreover, this decrease should obey a simple mathematical law, the *inverse-square law*. This means that the force acting on a planet at, say, twice the distance is reduced to one-quarter, the force acting on a planet at three times the distance is reduced to one-ninth, and so on. From this discovery—which was so beautiful that it surely concealed a deep truth about the world—it was a short step to the realization that it must be the Sun that causes the force in the first place. The Sun attracts the planet, but the attraction becomes weaker if the planet is farther away. It was a very appealing idea, and Newton took a giant intellectual leap: he assumed that the same kind of attractive force must exist between any two bodies whatsoever, anywhere in the universe.

And now, having "induced" the law for the force, Newton could bring the argument full circle by *deducing* the geometry of planetary motion. He *solved* the equations given by his laws of motion and gravity for a system of two mutually attracting bodies that obeyed his inverse-square law; in those days, "solved" meant finding a mathematical formula for their motion. The formula implied that they must move in ellipses about their common center of mass.

As Mars moves around the Sun in a giant ellipse, the Sun moves in an ellipse so tiny that its motion goes undetected. Indeed, the Sun is so massive compared to Mars that the mutual center of mass lies beneath the Sun's surface, which explains why Kepler thought that Mars moved in an ellipse around the stationary Sun.

However, when Newton and his sucessors tried to build on this success by solving the equation for a system of three or more bodies—such as Moon/Earth/Sun, or the entire Solar System—they ran into technical trouble; and they could get out of trouble only by changing the meaning of the word "solve." They failed to find any formulas that would solve the equations exactly, so they gave up looking for them. Instead, they tried to find ways to calculate approximate numbers. For example, around 1860 the French astronomer Charles-Eugène Delaunay filled an entire *book* with a single approximation to the motion of the Moon, as influenced by the gravitational attractions of the Earth and the Sun. It was an extremely accurate approximation—which is why it filled a book—and it took him twenty years to work it out. When it was subsequently checked, in 1970, using a symbolic-algebra computer program, the calculation took a mere twenty hours: only three mistakes were found in Delaunay's work, none serious.

The motion of the Moon/Earth/Sun system is said to be a three-body problem—for evident reasons. It is so unlike the nice, tidy two-body problem Newton solved that it might as well have been invented on another planet in another galaxy, or in another universe. The three-body problem asks for a solution for the equations that describe the motion of three masses under inverse-square-law gravity. Mathematicians tried to find such a solution for centuries but met with astonishingly little success beyond approximations, such as Delaunay's, which worked only for particular cases, like Moon/Earth/Sun. Even the so-called restricted three-body problem, in which one body has a mass so small that it can be considered to exert no force at all upon the other two, proved utterly intractable. It was the first serious hint that knowing the laws might not be enough to understand how a system behaves; that the crevasse between laws and behavior might not always be bridgeable.

Despite intensive effort, more than three centuries after Newton we still do not have a complete answer to the three-body problem. However, we finally know why the problem has been so hard to crack. The two-body problem is "integrable"—the laws of conservation of energy and momentum restrict solutions so much that they are forced to take a simple mathematical form. In 1994, Zhihong Xia, of the Georgia Institute of Technology, proved what mathematicians had long suspected: that a system of three bodies is not integrable. Indeed, he did far more, by showing that such a system can exhibit a strange phenomenon known as

Arnold diffusion, first discovered by Vladimir Arnold, of Moscow State University, in 1964. Arnold diffusion produces an extremely slow, "random" drift in the relative orbital positions. This drift is not truly random: it is an example of the type of behavior now known as chaos—which can be described as apparently random behavior with purely deterministic causes.

Notice that this approach again changes the meaning of "solve." First that word meant "find a formula." Then its meaning changed to "find approximate numbers." Finally, it has in effect become "tell me what the solutions look like." In place of quantitative answers, we seek qualitative ones. In a sense, what is happening looks like a retreat: if it is too hard to find a formula, then try an approximation; if approximations aren't available, try a qualitative description. But it is wrong to see this development as a retreat, for what this change of meaning has taught us is that for questions like the three-body problem, no formulas *can* exist. We can prove that there are qualitative aspects to the solution that a formula cannot capture. The search for a formula in such questions was a hunt for a mare's nest.

Why did people want a formula in the first place? Because in the early days of dynamics, that was the only way to work out what kind of motion would occur. Later, the same information could be deduced from approximations. Nowadays, it can be obtained from theories that deal directly and precisely with the main qualitative aspects of the motion. As we will see in the next few chapters, this move toward an explicitly qualitative theory is not a retreat but a major advance. For the first time, we are starting to understand nature's patterns in their own terms.

CHAPTER 5

FROM VIOLINS TO VIDEOS

It has become conventional, as I have noted, to separate mathematics into two distinct subdisciplines labeled pure mathematics and applied mathematics. This is a separation that would have baffled the great mathematicians of classical times. Carl Friedrich Gauss, for example, was happiest in the ivory tower of number theory, where he delighted in abstract numerical patterns simply because they were beautiful and challenging. He called number theory "the queen of mathematics," and the poetic idea that queens are delicate beauties who do not sully their hands with anything useful was not far from his mind. However, he also calculated the orbit of Ceres, the first asteroid to be discovered. Soon after its discovery, Ceres passed behind the Sun, as seen from Earth, and could no longer be observed. Unless its orbit could be calculated accurately, astronomers would not be able to find it when it again became visible, months later. But the number of observations of the asteroid was so small that the standard methods for calculating orbits could not provide the required level of accuracy. So Gauss made several major innovations, some of which remain in use to this day. It was a virtuoso performance, and it made his public reputation. Nor was that his only practical application of his subject: among other things, he was also responsible for major developments in surveying, telegraphy, and the understanding of magnetism.

In Gauss's time, it was possible for one person to have a fairly good grasp of the whole of mathematics. But because all of the classical branches of science have grown so vast that no single mind can likely encompass even one of them, we now live in an age of specialists. The organizational aspects of mathematics function more tidily if people specialize either in the theoretical areas of the subject or its practical ones. Because most people feel happier working in one or the other of these two styles, individual preferences tend to reinforce this distinction. Unfortunately, it is then very tempting for the outside world to assume that the only useful part of mathematics is applied mathematics; after all, that is what the name seems to imply. This assumption is correct when it comes to established mathematical techniques: anything really useful inevitably ends up being considered "applied," no matter what it origins may have been. But it gives a very distorted view of the origins of new mathematics of practical importance. Good ideas are rare, but they come at least as often from imaginative dreams about the internal structure of mathematics as they do from attempts to solve a specific, practical problem. This chapter deals with a case history of just such a development, whose most powerful application is television—an invention that arguably has changed our world more than any other. It is a story in which the pure and applied aspects of mathematics combine to yield something far more powerful and compelling than either could have produced alone. And it begins at the start of the sixteenth century, with the problem of the vibrating violin string. Although this may sound like a practical question, it was studied mainly as an exercise in the solution of differential equations; the work was not aimed at improving the quality of musical instruments.

Imagine an idealized violin string, stretched in a straight line between two fixed supports. If you pluck the string, pulling it away from the straight-line position and then letting go, what happens? As you pull it sideways, its elastic tension increases, which produces a force that pulls the string back toward its original position. When you let go, it begins to accelerate under the action of this force, obeying Newton's law of motion. However, when it returns to its initial position it is moving rapidly, because it has been accelerating the whole time—so it overshoots the straight line and keeps moving. Now the tension pulls in the opposite direction, slowing it down until it comes to a halt. Then the whole story starts over. If there is no friction, the string will vibrate from side to side forever.

That's a plausible verbal descripton; one of the tasks for a mathematical theory is to see whether this scenario really holds good, and if so, to work out the details, such as the shape that the string describes at any instant. It's a complex problem, because the same string can vibrate in many different ways, depending upon how it is plucked. The ancient Greeks knew this, because their experiments showed that a vibrating string can produce many different musical tones. Later generations realized that the pitch of the tone is determined by the frequency of vibration—the rate at which the string moves to and fro—so the Greek discovery tells us that the same string can vibrate at many different frequencies. Each frequency corresponds to a different configuration of the moving string, and the same string can take up many different shapes.

Strings vibrate much too fast for the naked eye to see any one instantaneous shape, but the Greeks found important evidence for the idea that a string can vibrate at many different frequencies. They showed that the pitch depends on the positions of the *nodes*—places along the length of the string which remain stationary. You can test this on a violin, banjo, or guitar. When the string is vibrating in its "fundamental" frequency—that is, with the lowest possible pitch—only the end points are at rest. If you place a finger against the center of the string, creating a node, and then pluck the string, it produces a note one octave higher. If you place your finger one-third of the way along the string, you actually create two nodes (the other being two-thirds of the way along), and this produces a yet higher note. The more nodes, the higher the frequency. In general, the number of nodes is an integer, and the nodes are equally spaced.

The corresponding vibrations are standing waves, meaning waves that move up and down but do not travel along the string. The size of the up-and-down movement is known as the amplitude of the wave, and this determines the tone's loudness. The waves are sinusoidal—shaped like a sine curve, a repetitive wavy line of rather elegant shape that arises in trigonometry.

In 1714, the English mathematician Brook Taylor published the fundamental vibrational frequency of a violin string in terms of its length, tension, and density. In 1746, the Frenchman Jean Le Rond d'Alembert showed that many vibrations of a violin string are not sinusoidal standing waves. In fact, he proved that the instantaneous shape of the wave can be anything you like. In 1748, in response to d'Alembert's work, the prolific Swiss mathematician Leonhard Euler worked out the "wave equation" for a string. In the spirit of Isaac Newton, this is a differential equation that governs the rate of change of the shape of the string. In fact it is a "partial differential equation," meaning that it involves not only rates of change relative to time but also rates of change relative to space—the direction along the string. It expresses in mathematical language the idea that the acceleration of each tiny segment of the string is proportional to the tensile forces acting upon that segment; so it is a consequence of Newton's law of motion.

Not only did Euler formulate the wave equation: he solved it. His solution can be described in words. First, deform the string into any shape you care to choose—a parabola, say, or a triangle, or a wiggly and irregular curve of your own devising. Then imagine that shape propagating along the string toward the right. Call this a rightward-traveling wave. Then turn the chosen shape upside down, and imagine it propagating the other way, to create a leftward-traveling wave. Finally, superpose these two waveforms. This process leads to all possible solutions of the wave equation in which the ends of the string remain fixed.

Almost immediately, Euler got into an argument with Daniel Bernoulli, whose family originally hailed from Antwerp but had moved to Germany and then Switzerland to escape religious persecution. Bernoulli also solved the wave equation, but by a totally different method. According to Bernoulli, the most general solution can be represented as a superposition of infinitely many sinusoidal standing waves. This apparent disagreement began a century-long controversy, eventually resolved by declaring both Euler and Bernoulli right. The reason that they are both right is that every periodically varying shape can be represented as a superposition of an infinite number of sine curves. Euler thought that his approach led to a greater variety of shapes, because he didn't recognize their periodicity. However, the mathematical analysis works with an infinitely long curve. Because the only part of the curve that matters is the part between the two endpoints, it can be repeated periodically along a very long string without any essential change. So Euler's worries were unfounded.

The upshot of all this work, then, is that the sinusoidal waves are the basic vibrational components. The totality of vibrations that can occur is given by forming all possible sums of finitely or infinitely many sinusoidal waves of all possible amplitudes. As Daniel Bernoulli had maintained all along, "all new curves given by d'Alembert and Euler are only combinations of the Taylor vibrations."

With the resolution of this controversy, the vibrations of a violin string ceased to be a mystery, and the mathematicians went hunting for bigger game. A violin string is a curve—a one-dimensional object—but objects with more dimensions can also vibrate. The most obvious musical instrument that employs a two-dimensional vibration is the drum, for a drumskin is a surface, not a straight line. So mathematicians turned their attention to drums, starting with Euler in 1759. Again he derived a wave equation, this one describing how the displacement of the drumskin in the vertical direction varies over time. Its physical interpretation is that the acceleration of a small piece of the drumskin is proportional to the average tension exerted on it by all nearby parts of the drumskin: symbolically, it looks much like the one-dimensional wave equation; but now there are spatial (second-order) rates of change in two independent directions, as well as the temporal rate of change.

Violin strings have fixed ends. This "boundary condition" has an important effect: it determines which solutions to the wave equation are physically meaningful for a violin string. In this whole subject, boundaries are absolutely crucial. Drums differ from violin strings not only in their dimensionality but in having a much more interesting boundary: the boundary of a drum is a closed curve, or circle. However, like the boundary of a string, the boundary of the drum is fixed: the rest of the drumskin can move, but its rim is firmly strapped down. This boundary condition restricts the possible motions of the drumskin. The isolated endpoints of a violin string are not as interesting and varied a boundary condition as a closed curve is; the true role of the boundary becomes apparent only in two or more dimensions.

As their understanding of the wave equation grew, the mathematicians of the eighteenth century learned to solve the wave equation for the motion of drums of various shapes. But now the wave equation began to move out of the musical domain to establish itself as an absolutely central feature of mathematical physics. It is probably the single most important mathematical formula ever devised—Einstein's famous relation between mass and energy notwithstanding. What happened was a dramatic instance of how mathematics can lay bare the hidden unity of nature. The same equation began to show up *everywhere*. It showed up in fluid dynamics, where it described the formation and motion of water waves. It showed up in the theory of sound, where it described the transmission of sound waves—vibrations of the air, in which its molecules become alternately compressed and separated. And then it showed up in the theories of electricity and magnetism, and changed human culture forever.

Electricity and magnetism have a long, complicated history, far more complex than that of the wave equation, involving accidental discoveries and key experiments as well as mathematical and physical theories. Their story begins with William Gilbert, physician to Elizabeth I, who described the Earth as a huge magnet and observed that electrically charged bodies can attract or repel each other. It continues with such people as Benjamin Franklin, who in 1752 proved that lightning is a form of electricity by flying a kite in a thunderstorm; Luigi Galvani, who noticed that electrical sparks caused a dead frog's leg muscles to contract; and Alessandro Volta, who invented the first battery. Throughout much of this early development, electricity and magnetism were seen as two quite distinct natural phenomena. The person who set their unification in train was the English physicist and chemist Michael Faraday. Faraday was employed at the Royal Institution in

London, and one of his jobs was to devise a weekly experiment to entertain its scientifically-minded members. This constant need for new ideas turned Faraday into one of the greatest experimental physicists of all time. He was especially fascinated by electricity and magnetism, because he knew that an electric current could create a magnetic force. He spent ten years trying to prove that, conversely, a magnet could produce an electric current, and in 1831 he succeeded. He had shown that magnetism and electricity were two different aspects of the same thing—electromagnetism. It is said that King William IV asked Faraday what use his scientific parlor tricks were, and received the reply "I do not know, Your Majesty, but I do know that one day you will tax them." In fact, practical uses soon followed, notably the electric motor (electricity creates magnetism creates motion) and the electrical generator (motion creates magnetism creates electricity). But Faraday also advanced the *theory* of electromagnetism. Not being a mathematician, he cast his ideas in physical imagery, of which the most important was the idea of a line of force. If you place a magnet under a sheet of paper and sprinkle iron filings on top, they will line up along well-defined curves. Faraday's interpretation of these curves was that the magnetic force did not act "at a distance" without any intervening medium; instead, it propagated through space along curved lines. The same went for electrical force.

Faraday was no mathematician, but his intellectual successor James Clerk Maxwell was. Maxwell expressed Faraday's ideas about lines of force in terms of mathematical equations for magnetic and electric fields—that is, distributions of magnetic and electrical charge throughout space. By 1864, he had refined his theory down to a system of four differential equations that related changes in the magnetic field to changes in the electric field. The equations are elegant, and reveal a curious symmetry between electricity and magnetism, each affecting the other in a similar manner.

It is here, in the elegant symbolism of Maxwell's equations, that humanity made the giant leap from violins to videos: a series of simple algebraic manipulations extracted the wave equation from Maxwell's equations—which implied the existence of electromagnetic waves. Moreover, the wave equation implied that these electromagnetic waves traveled with the speed of light. One immediate deduction was that light itself is an electromagnetic wave—after all, the most obvious thing that travels at the speed of light is light. But just as the violin string can vibrate at many frequencies, so—according to the wave equation—can the electromagnetic field. For waves that are visible to the human eye, it turns out that frequency corresponds to color. Strings with different frequencies produce different sounds; visible electromagnetic waves with different frequencies produce different colors. When the frequency is outside the visible range, the waves are not light waves but something else.

What? When Maxwell proposed his equations, nobody knew. In any case, all this was pure surmise, based on the assumption that Maxwell's equations really do apply to the physical world. His equations needed to be tested before these waves could be accepted as real. Maxwell's ideas found some favor in Britain, but they were almost totally ignored abroad until 1886, when the German physicist Heinrich Hertz generated electromagnetic waves—at the frequency that we now call radio—and detected them experimentally. The final episode of the saga was supplied by Guglielmo Marconi, who successfully carried out the first wireless telegraphy in 1895 and transmitted and received the first transatlantic radio signals in 1901.

The rest, as they say, is history. With it came radar, television, and videotape.

Of course, this is just a sketch of a lengthy and intricate interaction between mathematics, physics, engineering, and finance. No single person can claim credit for the invention of radio, neither can any single subject. It is conceivable that, had the mathematicians not already known a lot more about the wave equation, Maxwell or his successors would have worked out what it implied anyway. But ideas have to attain a critical mass before they explode, and no innovator has the time or the imagination to create the tools to make the tools to make the tools that

. . . even if they are intellectual tools. The plain fact is that there is a clear historical thread beginning with violins and ending with videos. Maybe on another planet things would have happened differently; but that's how they happened on ours.

And maybe on another planet things would not have happened differently—well, not very differently. Maxwell's wave equation is extremely complicated: it describes variations in both the electrical and magnetic fields simultaneously, in three-dimensional space. The violin-string equation is far simpler, with variation in just one quantity—position—along a one-dimensional line. Now, mathematical discovery generally proceeds from the simple to the complex. In the absence of experience with simple systems such as vibrating strings, a "goal-oriented" attack on the problem of wireless telegraphy (sending messages without wires, which is where that slightly old-fashioned name comes from) would have stood no more chance of success than an attack on antigravity or faster-than-light drives would do today. *Nobody would know where to start.*

Of course, violins are accidents of human culture—indeed, of European culture. But vibrations of a linear object are universal—they arise all over the place in one guise or another. Among the arachnid aliens of Betelgeuse II, it might perhaps have been the vibrations of a thread in a spiderweb, created by a struggling insect, that led to the discovery of electromagnetic waves. But it takes some clear train of thought to devise the particular sequence of experiments that led Heinrich Hertz to his epic discovery, and the train of thought has to start with something simple. And it is mathematics that reveals the simplicities of nature, and permits us to generalize from simple examples to the complexities of the real world. It took many people from many different areas of human activity to turn a mathematical insight into a useful product. But the next time you go jogging wearing a Walkman, or switch on your TV, or watch a videotape, pause for a few seconds to remember that without mathematicians none of these marvels would ever have been invented.

CHAPTER 6

BROKEN SYMMETRY

Something in the human mind is attracted to symmetry. Symmetry appeals to our visual sense, and thereby plays a role in our sense of beauty. However, perfect symmetry is repetitive and predictable, and our minds also like surprises, so we often consider imperfect symmetry to be more beautiful than exact mathematical symmetry. Nature, too, seems to be attracted to symmetry, for many of the most striking patterns in the natural world are symmetric. And nature also seems to be dissatisifed with too much symmetry, for nearly all the symmetric patterns in nature are less symmetric than the cause that give rise to them.

This may seem a strange thing to say; you may recall that the great physicist Pierre Curie, who with his wife, Marie, discovered radioactivity, stated the general principle that "effects are as symmetric as their causes." However, the world is full of effects that are *not* as symmetric as their causes, and the reason for this is a phenomenon known as "spontaneous symmetry breaking."

Symmetry is a mathematical concept as well as an aesthetic one, and it allows us to classify different types of regular pattern and distinguish between them. Symmetry breaking is a more dynamic idea, describing changes in pattern. Before we can understand where nature's patterns come from and how they can change, we must find a language in which to describe what they are.

What *is* symmetry?

Let's work our way to the general from the particular. One of the most familiar symmetric forms is the one inside which you spend your life. The human body is "bilaterally symmetric," meaning that its left half is (near enough) the same as its right half. As noted, the bilateral symmetry of the human form is only approximate: the heart is not central, nor are the two sides of the face identical. But the overall form is very close to one that has perfect symmetry, and in order to describe the mathematics of symmetry we can imagine an idealized human figure whose left side is exactly the same as its right side. But exactly the *same*? Not entirely. The two side of the figure occupy different regions of space; moreover, the left side is a reversal of the right—its mirror image.

As soon as we use words like "image," we are already thinking of how one shape corresponds to the other—of how you might move one shape to bring it into coincidence with the other. Bilateral symmetry means that if you reflect the left half in a mirror, then you obtain the right half. Reflection is a mathematical concept, but it is not a shape, a number, or a formula. It is a *transformation*—that is, a rule for moving things around.

There are many possible transformations, but most are not symmetries. To relate the halves correctly, the mirror must be placed on the *symmetry axis*, which divides the figure into its two related halves. Reflection then leaves the human form *invariant*—that is, unchanged in appearance. So we have found a precise mathematical characterization of bilateral symmetry—a shape is bilaterally symmetric if it is invariant by reflection. More generally, a symmetry of an object or system is any transformation that leaves it invariant. This description is a wonderful example of what I earlier called the "thingification of processes": the process "move like *this*" becomes a thing—a symmetry. This simple but elegant characterization opens the door to an immense area of mathematics.

There are many different kinds of symmetry. The most important ones are reflections, rotations, and translations—or, less formally, flips, turns, and slides. If you take an object in the plane, pick it up, and flip it over onto its back, you get the same effect as if you had reflected it in a suitable mirror. To find where the mirror should go, choose some point on the original object and look at where that point ends up when the object is flipped. The mirror must go halfway between the point and its image, at right angles to the line that joins them (see figure 3). Reflections can also be carried out in three-dimensional space, but now the mirror is of a more familiar kind—namely, a flat surface.

To rotate an object in the plane, you choose a point,

called the center, and turn the object about that center, as a wheel turns about its hub. The number of degrees through which you turn the object determines the "size" of the rotation. For example, imagine a flower with four identical equally spaced petals. If you rotate the flower 90°, it looks exactly the same, so the transformation "rotate through a right angle" is a symmetry of the flower. Rotations can occur in three-dimensional space too, but now you have to choose a line, the axis, and spin objects on that axis as the Earth spins on its axis. Again, you can rotate objects through different angles about the same axis.

Translations are transformations that slide objects along without rotating them. Think of a tiled bathroom wall. If you take a tile and slide it horizontally just the right distance, it will fit on top of a neighboring tile. That distance is the width of a tile. If you slide it two widths of a tile, or three, or any whole number, it also fits the pattern. The same is true if you slide it in a vertical direction, or even if you use a combination of horizontal and vertical slides. In fact, you can do more than just sliding one tile—you can slide the entire pattern of tiles. Again, the pattern fits neatly on top of its original position only when you use a combination of horizontal and vertical slides through distances that are whole number multiples of the width of a tile.

Reflections capture symmetries in which the left half of a pattern is the same as the right half, like the human body.

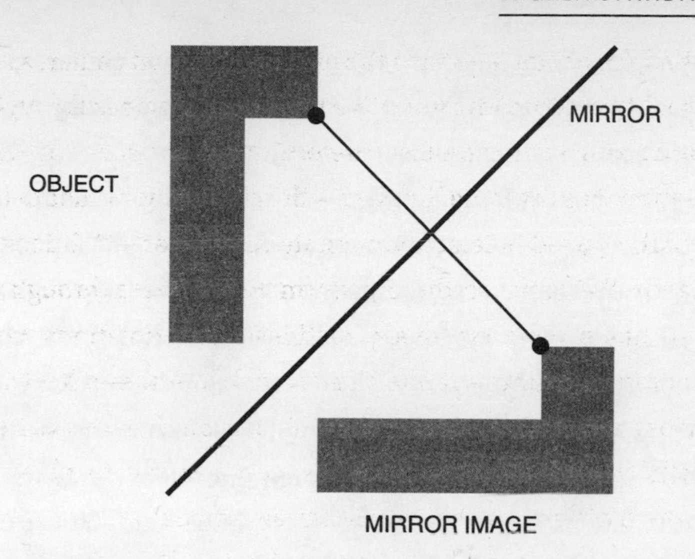

FIGURE 3.
Where is the mirror? Given an object and a mirror image of that object, choose any point of the object and the corresponding point of the image. Join them by a line. The mirror must be at right angles to the midpoint of that line.

Rotations capture symmetries in which the same units repeat around circles, like the petals of a flower. Translations capture symmetries in which units are repeated, like a regular array of tiles; the bees' honeycomb, with its hexagonal "tiles," is an excellent naturally occurring example.

Where do the symmetries of natural patterns come from? Think of a still pond, so flat that it can be thought of as a mathematical plane, and large enough that it might as well *be* a plane for all that the edges matter. Toss a pebble into the pond. You see patterns, ripples, circular waves seemingly moving outward away from the point of impact of the pebble. We've all seen this, and nobody is greatly surprised. After all, we *saw* the cause: it was the pebble. If you don't throw pebbles in, or anything else that might disturb the surface, then you won't get waves. All you'll get is a still, flat, planar pond.

Ripples on a pond are examples of broken symmetry. An ideal mathematical plane has a huge amount of symmetry: every part of it is identical to every other part. You can translate the plane through any distance in any direction, rotate it through any angle about any center, reflect it in any mirror line, and it still looks exactly the same. The pattern of circular ripples, in contrast, has less symmetry. It is symmetric only with respect to rotations about the point of impact of the pebble, and reflections in mirror lines that run through that point. No translations, no other rotations, no other reflections. The pebble breaks the symmetry of the plane, in the sense that after the pebble has disturbed the pond, many of its symmetries are lost. But *not all*, and that's why we see a pattern.

However, none of this is surprising, because of the pebble. In fact, since the impact of the pebble creates a special point, different from all the others, the symmetries of the ripples are exactly what you would expect. They are precisely the symmetries that do not move that special point. So the symmetry of the pond is not *spontaneously* broken when the ripples appear, because you can detect the stone that causes the translational symmetries to be lost.

You would be more surprised—a lot more surprised—if a perfectly flat pond suddenly developed a series of concentric circular ripples without there being any obvious cause. You would imagine that perhaps a fish beneath the surface had disturbed it, or that something had fallen in and you had not seen it because it was moving too fast. So strong is the ingrained assumption that patterns must have evident causes that when in 1958 the Russian chemist B.P. Belousov discovered a chemical reaction that spontaneously formed patterns, apparently out of nothing, his colleagues refused to believe him. They assumed that he had made a mistake. They didn't bother checking his work: he was so obviously wrong that checking his work would be a waste of time.

Which was a pity, because he was right.

The particular pattern that Belousov discovered existed not in space but in time: his reaction oscillated through a periodic sequence of chemical changes. By 1963, another Russian chemist, A. M. Zhabotinskii, had modified Belousov's reaction so that it formed patterns in space as well. In their honor, any similar chemical reaction is given the generic name "Belousov-Zhabotinskii [or B-Z] reaction." The chemicals used nowadays are different and simpler, thanks to some refinements made by the British reproductive biologist Jack Cohen and the American mathematical biologist Arthur Winfree, and the experiment is now so simple that it can be done by anyone with access to the necessary chemicals. These are slightly esoteric, but there are only four of them.*

In the absence of the appropriate apparatus, I'll tell you what happens if you do the experiment. The chemicals are all liquids: you mix them together in the right order and pour them into a flat dish. The mixture turns blue, then red: let it stand for a while. For ten or sometimes even twenty minutes, nothing happens; it's just like gazing at a featureless flat pond—except that it is the color of the liquid that is featureless, a uniform red. This uniformity is not surprising; after all, you blended the liquids. Then you notice a few tiny blue spots appearing—and that is a surprise. They spread, forming circular blue disks. Inside each disk, a red spot appears, turning the disk into a blue ring with a red center. Both the blue ring and the red disk grow, and when the red disk gets big enough, a blue spot appears inside it. The process continues, forming an ever-growing series of "target patterns"—concentric rings of red and blue. These target patterns have exactly the same symmetries as the rings of ripples on a pond; but this time you can't see any pebble. It is a strange and mysterious process in which

* The precise recipe is given in the Notes to *The Collapse of Chaos*, by Jack Cohen and Ian Stewart.

pattern—order—appears to arise of its own accord from the disordered, randomly mixed liquid. No wonder the chemists didn't believe Belousov.

But that's not the end of the B-Z reaction's party tricks. If you tilt the dish slightly and then put it back where it was, or dip a hot wire into it, you can break the rings and turn them into rotating red and blue spirals. If Belousov had claimed *that*, you would have seen steam coming out of his colleagues' ears.

This kind of behavior is not just a chemical conjuring trick. The regular beating of your heart relies on exactly the same patterns, but in that case they are patterns in waves of electrical activity. Your heart is not just a lump of undifferentiated muscle tissue, and it doesn't automatically contract all at once. Instead, it is a composed of millions of tiny muscle fibers, each one of them a single cell. The fibers contract in response to electrical and chemical signals, and they pass those signals on to their neighbors. The problem is to make sure that they all contract roughly in synchrony, so that the heart beats as a whole. To achieve the necessary degree of synchronization, your brain sends electrical signals to your heart. These signals trigger electrical changes in some of the muscle fibers, which then affect the muscle fibers next to them—so that ripples of activity spread, just like the ripples on a pond or the blue disks in the B-Z reaction. As long as the waves form complete rings, the heart's muscle fibers contract in synchrony and the heart beats normally. But if the waves become spirals—as they can do in diseased hearts—the result is an incoherent set of local contractions, and the heart fibrillates. If fibrillation goes unchecked for more than a few minutes, it results in death. So every single one of us has a vested interest in circular and spiral wave patterns.

However in the heart, as in the pond, we can see a specific cause for the wave patterns: the signals from the brain. In the B-Z reaction, we cannot: the symmetry breaks spontaneously—"of its own accord"—without any external stimulus. The term "spontaneous" does not imply that there is *no* cause, however: it indicates that the cause can be as tiny and as insignificant as you please. Mathematically, the crucial point is that the uniform distribution of chemicals—the featureless red liquid—is unstable. If the chemicals cease to be equally mixed, then the delicate balance that keeps the solution red is upset, and the resulting chemical changes trigger the formation of a blue spot. From that moment on, the whole process becomes much more comprehensible, because now the blue spot acts like a chemical "pebble," creating sequential ripples of chemical activity. But—at least, as far as the mathematics goes—the imperfection in the symmetry of the liquid which triggers the blue spot can be vanishingly small, provided it is not zero. In a real liquid, there are always tiny bits of dust, or bubbles—or even just molecules undergoing the vibrations we call "heat"—to disturb the perfect symmetry. That's all it takes. An infinitesimal cause produces a large-scale effect, and that effect is a symmetric pattern.

Nature's symmetries can be found on every scale, from the structure of subatomic particles to that of the entire universe. Many chemical molecules are symmetric. The methane molecule is a tetrahedron—a triangular-sided pyramid—with one carbon atom at its center and four hydrogen atoms at its corners. Benzene has the sixfold symmetry of a regular hexagon. The fashionable molecule buckminsterfullerene is a truncated icosahedral cage of sixty carbon atoms. (An icosahedron is a regular solid with twenty triangular faces; "truncated" means that the corners are cut off.) Its symmetry lends it a remarkable stability, which has opened up new possibilities for organic chemistry.

On slightly larger scale than molecules, we find symmetries in cellular structure; at the heart of cellular replication lies a tiny piece of mechanical engineering. Deep within each living cell, there is a rather shapeless structure known as the centrosome, which sprouts long thin microtubules, basic components of the cell's internal "skeleton," like a diminutive sea urchin. Centrosomes were first discovered in 1887 and play an important role in organizing cell division. However, in one respect the structure of the centrosome is astonishingly symmetric. Inside it are two structures, known as centrioles, positioned at right angles to each other. Each centriole is cylindrical, made from twenty-seven microtubules fused together along their lengths in threes, and arranged with perfect ninefold symmetry. The microtubules themselves also have an astonishingly regular symmetric form. They are hollow tubes, made from a perfect regular checkerboard pattern of units that contain two distinct proteins, alpha- and beta-tubulin. One day, perhaps, we will understand *why* nature chose these symmetric forms. But it is amazing to see symmetric structures at the core of a living cell.

Viruses are often symmetric, too, the commonest shapes being helices and icosahedrons. The helix is the form of the influenza virus, for instance. Nature prefers the icosahedron above all other viral forms: examples include herpes, chicken-pox, human wart, canine infectious hepatitis, turnip yellow mosaic, adenovirus, and many others. The adenovirus is another striking example of the artistry of molecular engineering. It is made from 252 virtually identical subunits, with 21 of them, fitted together like billiard balls before the break, making up each triangular face. (Subunits along the edges lie on more than one face and corner units lie on three, which is why 20 × 21 is not equal to 252.)

Nature exhibits symmetries on larger scales, too. A developing frog embryo begins life as a spherical cell, then loses symmetry step by step as it divides, until it has become a blastula, thousands of tiny cells whose overall form is again spherical. Then the blastula begins to engulf part of itself, in the process known as gastrulation. During the early stages of this collapse, the embryo has rotational symmetry about an axis, whose position is often determined by the initial distribution of yolk in the egg, or sometimes by the point of sperm entry. Later this symmetry is broken, and only a single mirror symmetry is retained, leading to the bilateral symmetry of the adult.

Volcanoes are conical, stars are spherical, galaxies are spiral or elliptical. According to some cosmologists, the universe itself resembles nothing so much as a gigantic expanding ball. Any understanding of nature must include an understanding of these prevalent patterns. It must explain why they are so common, and why many different aspects of nature show the *same* patterns. Raindrops and stars are spheres, whirlpools and galaxies are spirals, honeycombs and the Giant's Causeway are arrays of hexagons. There has to be a general principle underlying such patterns; it is not enough just to study each example in isolation and explain it in terms of its own internal mechanisms.

Symmetry breaking is just such a principle.

But in order for symmetry to break, it has to be present to start with. At first this would seem to replace one problem of pattern formation with another: before we can explain the circular rings on the pond, in other words, we have to explain the pond. But there is a crucial difference between the rings and the pond. The symmetry of the pond is so extensive—every point on its surface being equivalent to every other—that we do not recognize it as being a pattern. Instead, we see it as bland uniformity. It is very easy to explain bland uniformity: it is what happens to systems when there is no reason for their component parts to differ from each other. It is, so to speak, nature's default option. If something is symmetric, its component features are replaceable or interchangeable. One corner of a square looks pretty much the same as any other, so we can interchange the corners without altering the square's appearance. One atom of hydrogen in methane looks pretty much like any other, so we can interchange those atoms. One region of stars in a galaxy looks pretty much like any other, so we can interchange parts of two different spiral arms without making an important difference.

In short, nature is symmetric because we live in a mass-produced universe—analogous to the surface of a pond. Every electron is exactly the same as every other electron, every proton is exactly the same as every other proton, every region of empty space is exactly the same as every other region of empty space, every instant of time is exactly the same as every other instant of time. And not only are the structure of space, time, and matter the same everywhere: so are the laws that govern them. Albert Einstein made such "invariance principles" the cornerstone of his approach to physics; he based his thinking on the idea that no particular point in spacetime is special. Among other things, this led him to the principle of relativity, one of the greatest physical discoveries ever made.

This is all very well, but it produces a deep paradox. If the laws of physics are the same at all places and at all times, why is there any "interesting" structure in the universe at all? Should it not be homogeneous and changeless? If every place in the universe were interchangeable with every other place, then all places would be indistinguishable; and the same would hold for all times. But they are not. The problem is, if anything, made worse by the cosmological theory that the universe began as a single point, which exploded from nothingness billions of years ago in the big bang. At the instant of the universe's formation, all places and all times were not only indistinguishable but identical. So why are they different now?

The answer is the failure of Curie's Principle, noted at the start of this chapter. Unless that principle is hedged around with some very subtle caveats about arbitrarily tiny causes, it offers a misleading intuition about how a symmetric system should behave. Its prediction that adult frogs should be bilaterally symmetric (because embryonic frogs are bilaterally symmetric, and according to Curie's Principle the symmetry cannot change) appears at first sight to be a great success; but the same argument applied at the spherical blastula stage predicts with equal force that an adult frog should be a sphere.

A much better principle is the exact opposite, the

principle of spontaneous symmetry breaking. Symmetric causes often produce *less* symmetric effects. The evolving universe can break the initial symmetries of the big bang. The spherical blastula can develop into the bilateral frog. The 252 perfectly interchangeable units of adenovirus can arrange themselves into an icosahedron—an arrangement in which some units will occupy special places, such as corners. A set of twenty-seven perfectly ordinary microtubules can get together to create a centriole.

Fine, but why patterns? Why not a structureless mess, in which *all* symmetries are broken? One of the strongest threads that runs through every study ever made of symmetry breaking is that the mathematics does not work this way. Symmetries break reluctantly. There is so much symmetry lying around in our mass-produced universe that there is seldom a good reason to break all of it. So rather a lot survives. Even those symmetries that do get broken are still present, in a sense, but now as potential rather than actual form. For example, when the 252 units of the adenovirus began to link up, any one of them could have ended up in a particular corner. In that sense, they are interchangeable. But only one of them actually *does* end up there, and in that sense the symmetry is broken: they are no longer fully interchangeable. But some of the symmetry remains, and we see an icosahedron.

In this view, the symmetries we observe in nature are broken traces of the grand, universal symmetries of our mass-produced universe. *Potentially* the universe could exist in any of a huge symmetric system of possible states, but actually it must select one of them. In so doing, it must trade some of its actual symmetry for unobservable, potential symmetry. But some of the actual symmetry may remain, and when it does we observe a pattern. Most of nature's symmetric patterns arise out of some version of this general mechanism.

In a negative sort of way, this rehabilitates Curie's Principle: if we permit tiny asymmetric disturbances, which can trigger an instability of the fully symmetric state, then our mathematical system is no longer perfectly symmetric. But the important point is that the tinest departure from symmetry in the cause can lead to a total loss of symmetry in the resulting effect—and there are *always* tiny departures. That makes Curie's principle useless for the prediction of symmetries. It is much more informative to model a real system after one with perfect symmetry, but to remember that such a model has many possible states, only one of which will be realized in practice. Small disturbances cause the real system to select states from the range available to the idealized perfect system. Today this approach to the behavior of symmetric systems provides one of the main sources of understanding of the general principles of pattern formation.

In particular, the mathematics of symmetry breaking unifies what at first sight appear to be very disparate phenomena. For example, think about the patterns in sand dunes mentioned in chapter 1. The desert can be modeled as a flat plane of sandy particles, the wind can be modeled as a fluid flowing across the plane. By thinking about the symmetries of such a system, and how they can break, many of the observed patterns of dunes can be deduced. For example, suppose the wind blows steadily in a fixed direction, so that the whole system is invariant under translations parallel to the wind. One way to break these translational symmetries is to create a periodic pattern of parallel stripes, at right angles to the wind direction. But this is the pattern that geologists call transverse dunes. If the pattern also becomes periodic in the direction along the stripes, then more symmetry breaks, and the wavy barchanoid ridges appear. And so on.

However, the mathematical principles of symmetry-breaking do not just work for sand dunes. They work for any system with the same symmetries—anything that flows across a planar surface creating patterns. You can apply the same basic model to a muddy river flowing across a coastal plain and depositing sediment, or the waters of a shallow sea in ebb and flow across the seabed—phenomena important in geology, because millions of years later the patterns that result have been frozen into the rock that the sandy seabed and the muddy delta became. The list of patterns is identical to that for dunes.

Or the fluid might be a liquid crystal, as found in digital-watch displays, which consist of a lot of long thin molecules that arrange themselves in pattens under the influence of a magnetic or electric field. Again, you find the same patterns. Or there might not be a fluid at all: maybe what moves is a chemical, diffusing through tissue and laying down genetic instructions for patterns on the skin of a developing animal. Now the analogue of transverse dunes is the stripes of a tiger or a zebra, and that of barchanoid ridges is the spots on a leopard or a hyena.

The same abstract mathematics; different physical and biological realizations. Mathematics is the ultimate in technology transfer—but with mental technology, ways of thinking, being transferred, rather than machines. This universality of symmetry breaking explains why living systems and nonliving ones have many patterns in common. Life itself is a process of symmetry creation—of replication; the universe of biology is just as mass-produced as the universe of physics, and the organic world therefore exhibits many of the patterns found in the inorganic world. The most obvious symmetries of living organisms are those of form—icosahedral viruses, the spiral shell of *Nautilus*, the helical horns of gazelles, the remarkable rotational symmetries of starfish and jellyfish and flowers. But symmetries in the living world go beyond form into behavior—and not just the symmetric rhythms of locomotion I mentioned earlier. The territories of fish in

Lake Huron are arranged just like the cells in a honeycomb—and for the same reasons. The territories, like the bee grubs, cannot all be in the same place—which is what perfect symmetry would imply. Instead, they pack themselves as tightly as they can without one being different from another, and the behavioral constraint by itself produces a hexagonally symmetric tiling. And that resembles yet another striking instance of mathematical technology transfer, for the same symmetry breaking mechanism arranges the atoms of a crystal into a regular lattice—a physical process that ultimately supports Kepler's theory of the snowflake.

One of the more puzzling types of symmetry in nature is mirror symmetry, symmetry with respect to a reflection. Mirror symmetries of three-dimensional objects cannot be realized by turning the objects in space—a left shoe cannot be turned into a right shoe by rotating it. However, the laws of physics are very nearly mirror-symmetric, the exceptions being certain interactions of subatomic particles. As a result, any molecule that is not mirror-symmetric potentially exists in two different forms—left- and right-handed, so to speak. On Earth, life has selected a particular molecular handedness: for example, for amino acids. Where does this particular handedness of terrestrial life come from? It could have been just an accident—primeval chance propagated by the mass-production techniques of replication. If so, we might imagine that on some distant planet, creatures exist whose molecules are mirror images of ours. On the other hand, there may be a deep reason for life everywhere to choose the same direction. Physicists currently recognize four fundamental forces in nature: gravity, electromagnetism, and the strong and weak nuclear interactions. It is known that the weak force violates mirror symmetry—that is, it behaves differently in left- or right-handed versions of the same physical problem. As the Austrian-born physicist Wolfgang Pauli put it, "The Lord is a weak left-hander." One remarkable consequence of this violation of mirror symmetry is the fact that the energy levels of molecules and that of their mirror images are not exactly equal. The effect is extremely small: the difference in energy levels between one particular amino acid and its mirror image is roughly one part in 10^{17}. This may seem very tiny—but we saw that symmetry breaking requires only a very tiny disturbance. In general, lower-energy forms of molecules should be favored in nature. For this amino acid, it can be calculated that with 98% probability the lower energy form will become dominant within a period of about a hundred thousand years. And indeed, the version of this amino acid which is found in living organisms is the lower-energy one.

In chapter 5, I mentioned the curious symmetry of Maxwell's equations relating electricity and magnetism. Roughly speaking, if you interchange all the symbols for the electric field with those for the magnetic field, you re-create the same equations. This symmetry lies behind Maxwell's unification of electrical and magnetic forces into a single electromagnetic force. There is an analogous symmetry—though an imperfect one—in the equations for the four basic forces of nature, suggesting an even grander unification: that all four forces are different aspects of the same thing. Physicists have already achieved a unification of the weak and electromagnetic forces. According to current theories, all four fundamental forces should become unified—that is, symmetrically related—at the very high energy levels prevailing in the early universe. This symmetry of the early universe is broken in our own universe. In short, there is an ideal mathematical universe in which all of the fundamental forces are related in a perfectly symmetric manner—but we don't live in it.

That means that our universe could have been different; it could have been any of the other universes that, potentially, could arise by breaking symmetry in a different way. That's quite a thought. But there is an even more intriguing thought: the same basic method of pattern formation, the same mechanism of symmetry breaking in a mass-produced universe, governs the cosmos, the atom, and us.

CHAPTER 7

THE RHYTHM OF LIFE

Nature is nothing if not rhythmic, and its rhythms are many and varied. Our hearts and lungs follow rhythmic cycles whose timing is adapted to our body's needs. Many of nature's rhythms are like the heartbeat: they take care of themselves, running "in the background." Others are like breathing: there is a simple "default" pattern that operates as long as nothing unusual is happening, but there is also a more sophisticated control mechanism that can kick in when necessary and adapt those rhythms to immediate needs. Controllable rhythms of this kind are particularly common—and particularly interesting—in locomotion. In legged animals, the default patterns of motion that occur when conscious control is not operating are called gaits.

Until the development of high-speed photography, it was virtually impossible to find out exactly how an animal's legs moved as it ran or galloped: the motion is too fast for the human eye to discern. Legend has it that the photographic technique grew out of a bet on a horse. In the 1870s, the railroad tycoon Leland Stanford bet twenty-five thousand dollars that at some times a trotting horse has all four feet completely off the ground. To settle the issue, a photographer, who was born Edward Muggeridge but changed his name to Eadweard Muybridge, photographed the different phases of the gait of the horse, by placing a line of cameras with tripwires for the horse to trot past. Stanford, it is said, won his bet. Whatever the truth of the story, we do know that Muybridge went on to pioneer the scientific study of gaits. He also adapted a mechanical device known as the zoetrope to display them as "moving pictures," a road that in short order led to Hollywood. So Muybridge founded both a science and an art.

Most of this chapter is about gait analysis, a branch of mathematical biology that grew up around the questions "How do animals move?" and "Why do they move like that?" To introduce a little more variety, the rest is about rhythmic patterns that occur in entire animal populations, one dramatic example being the synchronized flashing of some species of fireflies, which is seen in some regions of the Far East, including Thailand. Although biological interactions that take place in individual animals are very different from those that take place in populations of animals, there is an underlying mathematical unity, and one of the messages of this chapter is that the same general mathematical concepts can apply on many different levels and to many different things. Nature respects this unity, and makes good use of it.

The organizing principle behind many such biological cycles is the mathematical concept of an oscillator—a unit whose natural dynamic causes it to repeat the same cycle of behavior over and over again. Biology hooks together huge "circuits" of oscillators, which interact with each other to create complex patterns of behavior. Such "coupled oscillator networks" are the unifying theme of this chapter.

Why do systems oscillate at all? The answer is that this is the simplest thing you can do if you don't want, or are not allowed, to remain still. Why does a caged tiger pace up and down? Its motion results from a combination of two constraints. First, it feels restless and does not wish to sit still. Second, it is confined within the cage and cannot simply disappear over the nearest hill. The simplest thing you can do when you have to move but can't escape altogether is to oscillate. Of course, there is nothing that forces the oscillation to repeat a regular rhythm; the tiger is free to follow an irregular path around the cage. But the simplest option—and therefore the one most likely to arise both in mathematics and in nature—is to find some series of motions that works, and repeat it over and over again. And that is what we mean by a periodic oscillation. In chapter 5, I described the vibration of a violin string. That, too, moves in a periodic oscillation, and it does so for the same reasons as the tiger. It can't remain stil because it has been plucked, and it can't get away altogether because its ends are pinned down and its total energy cannot increase.

Many oscillations arise out of steady states. As conditions change, a system that has a steady state may lose it and begin to wobble periodically. In 1942, the German mathematician Eberhard Hopf fond a general mathematical condition that guarantees such behavior: in his honor, this scenario is known as Hopf bifurcation. The idea is to approximate the dynamics of the original system in a particularly simple way, and to see whether a periodic wobble arises in this simplified system. Hopf proved that if the simplified system wobbles, then so does the original system. The great advantage of this method is that the mathematical calculations are carried out only for the simplified system, where they are relatively straightforward, whereas the result of those calculations tells us how the original system behaves. It is difficult to tackle the original system directly, and Hopf's approach sidesteps the difficulties in a very effective manner.

The word "bifurcation" is used because of a particular mental image of what is happening, in which the periodic oscillations "grow out from" the original steady state like a ripple on a pond growing out from its center. The physical interpretation of this mental picture is that the oscillations are very small to start with, and steadily become larger. The speed with which they grow is unimportant here.

For example, the sounds made by a clarinet depend on Hopf bifurcation. As the clarinetist blows air into the instrument, the reed—which was stationary—starts to vibrate. If the air flows gently, the vibration is small and produces a soft note. If the musician blows harder, the vibration grows and the note becomes louder. The important thing is that the musician does not have to blow in an oscillatory way (that is, in a rapid series of short puffs) to make the reed oscillate. This is typical of Hopf bifurcation: if the simplified system passes Hopf's mathematical test, then the real system will begin to oscillate of its own accord. In this case, the simplified system can be interpreted as a fictitious mathematical clarinet with a rather simple reed, although such an interpretation is not actually needed to carry out the calculations.

Hopf bifurcation can be seen as a special type of symmetry breaking. Unlike the examples of symmetry breaking described in the previous chapter, the symmetries that break relate not to space but to time. Time is a single variable, so mathematically it corresponds to a line—the time axis. There are only two types of line symmetry: translations and reflections. What does it mean for a system to be symmetric under time translation? It means that if you observe the motion of the system and then wait for some fixed interval and observe the motion of the system again, you will see exactly the same behavior. That is a description of periodic oscillations: if you wait for an interval equal to the period, you see exactly the same thing. So periodic oscillations have time-translation symmetry.

What about reflectional symmetries of time? Those correspond to reversing the direction in which time flows, a more subtle and philosophically difficult concept. Time reversal is peripheral to this chapter, but it is an extremely interesting question, which deserves to be discussed somewhere, so why not here? The *law* of motion is symmetric under time reversal. If you make a film of any "legal" physical motion (one that obeys the laws), and run the movie backward, what you see is also a legal motion. However, the legal motions common in our world often look bizarre when run backward. Raindrops falling from the sky to create puddles are an everyday sight; puddles that spit raindrops skyward and vanish are not. The source of the difference lies in the initial conditions. Most initial conditions break time-reversal symmetry. For example, suppose we decide to start with raindrops falling downward. This is not a time-symmetric state: its time reversal would have raindrops falling upward. Even though the laws are time-reversible, the motion they produce need not be, because once the time-reversal symmetry has been broken by the choice of initial conditions, it remains broken.

Back to the oscillators. I've now explained that periodic oscillations possess time-translation symmetry, but I haven't yet told you what symmetry is broken to create that pattern. The answer is "all time translations." A state that is invariant under these symmetries must look exactly the same at all instants of time—not just intervals of one period. That is, it must be a steady state. So when a system whose state is steady begins to oscillate periodically, its time-translational symmetries decrease from all translations to only translations by a fixed interval.

This all sounds rather theoretical. However, the realization that Hopf bifurcation is really a case of temporal symmetry breaking has led to an extensive theory of Hopf bifurcation in systems that have other symmetries as well—especially spatial ones. The mathematical machinery does not depend on particular interpretations and can easily work with several different kinds of symmetry at once. One of the success stories of this approach is a general classification of the patterns that typically set in when a symmetric network of oscillators undergoes a Hopf bifurcation, and one of the areas to which it has recently been applied is animal locomotion.

Two biologically distinct but mathematically similar types of oscillator are involved in locomotion. The most obvious oscillators are the animal's limbs, which can be thought of as mechanical systems—linked assemblies of bones, pivoting at the joints, pulled this way and that by contracting muscles. The main oscillators that concern us here, however, are to be found in the creature's nervous system, the neural circuitry that generates the rhythmic electrical signals that in turn stimulate and control the limbs' activity. Biologists call such a circuit a CPG, which stands for "central pattern generator." Correspondingly, a student of mine took to referring to a limb by the acronym

LEG, allegedly for "locomotive excitation generator." Animals have two, four, six, eight, or more LEGs, but we know very little directly about the CPGs that control them, for reasons I shall shortly explain. A lot of what we do know has been arrived at by working backward—or forward, if you like—from mathematical models.

Some animals possess only one gait—only one rhythmic default pattern for moving their limbs. The elephant, for example, can only walk. When it wants to move faster, it ambles—but an amble is just a fast walk, and the patterns of leg movement are the same. Other animals possess many different gaits; take the horse, for example. At low speeds, horses walk; at higher speeds, they trot; and at top speed they gallop. Some insert yet another type of motion, a canter, between a trot and a gallop. The differences are fundamental: a trot isn't just a fast walk but a different kind of movement altogether.

In 1965 the American zoologist Milton Hildebrand noticed that most gaits possess a degree of symmetry. That is, when an animal bounds, say, both front legs move together and both back legs move together; the bounding gait preserves the animal's bilateral symmetry. Other symmetries are more subtle: for example, the left half of a camel may follow the same sequence of movements as the right, but half a period out of phase—that is, after a time delay equal to half the period. So the pace gait has its own characteristic symmetry: "reflect left and right, and shift the phase by half a period." You use exactly this type of symmetry breaking to move yourself around: despite your bilateral symmetry, you don't move both legs simultaneously! There's an obvious advantage to bipeds in not doing so: if they move both legs slowly at the same time they fall over.

The seven most common quadrupedal gaits are the trot, pace, bound, walk, rotary gallop, transverse gallop, and canter. In the trot, the legs are in effect linked in diagonal pairs. First the front left and back right hit the ground together, then the front right and back left. In the bound, the front legs hit the ground together, then the back legs. The pace links the movements fore and aft: the two left legs hit the ground, then the two right. The walk involves a more complex but equally rhythmic pattern: front left, back right, front right, back left, then repeat. In the rotary gallop, the front legs hit the ground almost together, but with the right (say) very slightly later than the left; then the back legs hit the ground almost together, but this time with the left very slightly later than the right. The transverse gallop is similar, but the sequence is reversed for the rear legs. The canter is even more curious: first front left, then back right, then the other two legs simultaneously. There is also a rarer gait, the pronk, in which all four legs move simultaneously.

The pronk is uncommon, outside the cartoons, but is sometimes seen in young deer. The pace is observed in camels, the bound in dogs; cheetahs use the rotary gallop to travel at top speed. Horses are among the more versatile quadrupeds, using the walk, trot, transverse gallop, and canter, depending on circumstances.

The ability to switch gait comes from the dynamics of CPGs. The basic idea behind CPG models is that the rhythms and the phase relations of animal gaits are determined by the natural oscillation pattens of relatively simple neural circuits. What might such a circuit look like? Trying to locate a specific piece of neural circuitry in an animal's body is like searching for a particular grain of sand in a desert: to map out the nervous system of all but the simplest of animals is well beyond the capabilities even of today's science. So we have to sneak up on the problem of CPG design in a less direct manner.

One approach is to work out the simplest type of circuit that might produce all the distinct but related symmetry patterns of gaits. At first, this looks like a tall order, and we might be forgiven if we tried to concoct some elaborate structure with switches that effected the change from one gait to another, like a car gearbox. But the theory of Hopf bifurcation tells us that there is a simpler and more natural way. It turns out that the symmetry patterns observed in gaits are strongly reminiscent of those found in symmetric networks of oscillators. Such networks naturally possess an entire repertoire of symmetry-breaking oscillations, and can switch between them in a natural manner. You don't need a complicated gearbox.

For example, a network representing the CPG of a biped requires only two identical oscillators, one for each leg. The mathematics shows that if two identical oscillators are coupled—connected so that the state of each affects that of the other—then there are precisely two typical oscillation patterns. One is the *in-phase* pattern, in which both oscillators behave identically. The other is the *out-of-phase* pattern, in which both oscillators behave identically except for a half-period phase difference. Suppose that this signal from the CPG is used to drive the muscles that control a biped's legs, by assigning one leg to each oscillator. The resulting gaits inherit the same two patterns. For the in-phase oscillation of the network, both legs move together: the animal performs a two-legged hopping motion, like a kangaroo. In contrast, the out-of-phase motion of the CPG produces a gait resembling the human walk. These two gaits are the ones most commonly observed in bipeds. (Bipeds can, of course, do other things; for example, they can hop along on one leg—but in that case they effectively turn themselves into one-legged animals.)

What about quadrupeds? The simplest model is now a system of four coupled oscillators—one for each leg. Now the mathematics predicts a greater variety of patterns, and nearly all of them correspond to observed gaits. The most symmetric gait, the pronk, corresponds to all four oscillators being synchronized—that is, to unbroken symmetry.

The next most symmetric gaits—the bound, the pace, and the trot—correspond to grouping the oscillators as two out-of-phase pairs: front/back, left/right, or diagonally. The walk is a circulating figure-eight pattern and, again, occurs naturally in the mathematics. The two kinds of gallop are more subtle. The rotary gallop is a mixture of pace and bound, and the transverse gallop is a mixture of bound and trot. The canter is even more subtle and not as well understood.

The theory extends readily to six-legged creatures such as insects. For example, the typical gait of a cockroach—and, indeed, of most insects—is the tripod, in which the middle leg on one side moves in phase with the front and back legs on the other side, and then the other three legs move together, half a period out of phase with the first set. This is one of the natural patterns for six oscillators connected in a ring.

The symmetry-breaking theory also explains how animals can change gait without having a gearbox: a single network of oscillators can adopt different patterns under different conditions. The possible transitions between gaits are also organized by symmetry. The faster the animal moves, the less symmetry its gait has: more speed breaks more symmetry. But an explanation of *why* they change gait requires more detailed information on physiology. In 1981, D. F. Hoyt and R. C. Taylor discovered that when horses are permitted to select their own speeds, depending on terrain, they choose which ever gait minimizes their oxygen consumption.

I've gone into quite a lot of detail about the mathematics of gaits because it is an unusual application of modern mathematical techniques in an area that at first sight seems totally unrelated. To end this chapter, I want to show you another application of the same general ideas, except that in this case it is biologically important that symmetry *not* be broken.

One of the most spectacular displays in the whole of nature occurs in Southeast Asia, where huge swarms of fireflies flash in synchrony. In his 1935 article "Synchronous Flashing of Fireflies" in the journal *Science*, the American biologist Hugh Smith provides a compelling description of the phenomenon:

> Imagine a tree thirty-five to forty feet high, apparently with a firefly on every leaf, and all the fireflies flashing in perfect unison at the rate of about three times in two seconds, the tree being in complete darkness between flashes. Imagine a tenth of a mile of river front with an unbroken line of mangrove trees with fireflies on every leaf flashing in synchronism, the insects on the trees at the ends of the line acting in perfect unison with those between. Then, if one's imagination is sufficiently vivid, he may form some conception of this amazing spectacle.

Why do flashes synchronize? In 1990, Renato Mirollo and Steven Strogatz showed that synchrony is the rule for mathematical models in which every firefly interacts with every other. Again, the idea is to model the insects as a population of oscillators coupled together—this time by visual signals. The chemical cycle used by each firefly to create a flash of light is represented as an oscillator. The population of fireflies is represented by a network of such oscillators with fully symmetric coupling—that is, each oscillator affects all of the others in exactly the same manner. The most unusual feature of this model, which was introduced by the American biologist Charles Peskin in 1975, is that the oscillators are pulse-coupled. That is, an oscillator affects its neighbors only at the instant when it creates a flash of light.

The mathematical difficulty is to disentangle all these interactions, so that their combined effect stands out clearly. Mirollo and Strogatz proved that no matter what the initial conditions are, eventually all the oscillators become synchronized. The proof is based on the idea of *absorption*, which happens when two oscillators with different phases "lock together" and thereafter stay in phase with each other. Because the coupling is fully symmetric, once a group of oscillators has locked together, it cannot unlock. A geometric and analytic proof shows that a sequence of these absorptions must occur, which eventually locks *all* the oscillators together.

The big message in both locomotion and synchronization is that nature's rhythms are often linked to symmetry, and that the patterns that occur can be classified mathematically by invoking the general principles of symmetry breaking. The principles of symmetry breaking do not answer every question about the natural world, but they do provide a unifying framework, and often suggest interesting new questions. In particular, they both pose and answer the question, Why these patterns but not others?

The lesser message is that mathematics can illuminate many aspects of nature that we do not normally think of as being mathematical. This is a message that goes back to the Scottish zoologist D'Arcy Thompson, whose classic but maverick book *On Growth and Form* set out, in 1917, an enormous variety of more or less plausible evidence for the role of mathematics in the generation of biological form and behavior. In an age when most biologists seem to think that the only interesting thing about an animal is its DNA sequence, it is a message that needs to be repeated, loudly and often.

CHAPTER 8

DO DICE PLAY GOD?

The intellectual legacy of Isaac Newton was a vision of the clockwork universe, set in motion at the instant of creation but thereafter running in prescribed grooves, like a well-oiled machine. It was an image of a totally deterministic world—one leaving no room for the operation of chance, one whose future was completely determined by its present. As the great mathematical astronomer Pierre-Simon de Laplace eloquently put it in 1812 in his *Analytic Theory of Probabilities*:

> An intellect which at any given moment knew all the forces that animate Nature and the mutual positions of the beings that comprise it, if this intellect were vast enough to submit its data to analysis, could condense into a single formula the movement of the greatest bodies of the universe and that of the lightest atom: for such an intellect nothing could be uncertain, and the future just like the past would be present before its eyes.

This same vision of a world whose future is totally predictable lies behind one of the most memorable incidents in Douglas Adams's 1979 science-fiction novel *The*

Hitchhiker's Guide to the Galaxy, in which the philosophers Majikthise and Vroomfondel instruct the supercomputer "Deep Thought" to calculate the answer to the Great Question in Life, the Universe and Everything. Aficionados will recall that after five million years the computer answered, "Forty-two," at which point the philosophers realized that while the answer was clear and precise, the question had not been. Similarly, the fault in Laplace's vision lies not in his answer—that the universe is in principle predictable, which is an accurate statement of a particular mathematical feature of Newton's law of motion—but in his interpretation of that fact, which is a serious misunderstanding based on asking the wrong question. By asking a more appropriate question, mathematicians and physicists have now come to understand that determinism and predictability are not synonymous.

In our daily lives, we encounter innumerable cases where Laplacian determinism seems to be a highly inappropriate model. We walk safely down steps a thousand times, until one day we turn our ankle and break it. We go to a tennis match, and it is rained off by an unexpected thunderstorm. We place a bet on the favorite in a horse race, and it falls at the last fence when it is six lengths ahead of the field. It's not so much a universe in which—as Albert Einstein memorably refused to believe—God plays dice: it seems more a universe in which dice play God.

Is our world deterministic, as Laplace claimed, or is it

governed by chance, as it so often seems to be? And if Laplace is really right, why does so much of our experience indicate that he is wrong? One of the most exciting new areas of mathematics, nonlinear dynamics—popularly known as chaos theory—claims to have many of the answers. Whether or not it does, it is certainly creating a revolution in the way we think about order and disorder, law and chance, predictability and randomness.

According to modern physics, nature is ruled by chance on its smallest scales of space and time. For instance, whether a radioactive atom—of uranium, say—does or does not decay at any given instant is purely a matter of chance. There is no physical difference whatsoever between a uranium atom that is about to decay and one that is not about to decay. None. Absolutely none.

There are at least two contexts in which to discuss these issues: quantum mechanics and classical mechanics. Most of this chapter is about classical mechanics, but for a moment let us consider the quantum-mechanical context. It was this view of quantum indeterminacy that prompted Einstein's famous statement (in a letter to his colleague Max Born) that "you believe in a God who plays dice, and I in complete law and order." To my mind, there is something distinctly fishy about the orthodox physical view of quantum indeterminacy, and I appear not to be alone, because, increasingly, many physicists are beginning to wonder whether Einstein was right all along and something is missing from conventional quantum mechanics—perhaps "hidden variables," whose values tell an atom when to decay. (I hasten to add that this is not the conventional view.) One of the best known of them, the Princeton physicist David Bohm, devised a modification of quantum mechanics that is fully deterministic but entirely consistent with all the puzzling phenomena that have been used to support the conventional view of quantum indeterminacy. Bohm's ideas have problems of their own, in particular a kind of "action at a distance" that is no less disturbing than quantum indeterminacy.

However, even if quantum mechanics is correct about indeterminacy on the smallest scales, on macroscopic scales of space and time the universe obeys deterministic laws. This results from an effect called *decoherence*, which causes sufficiently large quantum systems to lose nearly all of their indeterminacy and behave much more like Newtonian systems. In effect, this reinstates classical mechanics for most human-scale purposes. Horses, the weather, and Einstein's celebrated dice are not unpredictable because of quantum mechanics. On the contrary, they are unpredictable within a Newtonian model, too. This is perhaps not so surprising when it comes to horses—living creatures have their own hidden variables, such as what kind of hay they had for breakfast. But it was definitely a surprise to those meteorologists who had been developing massive computer simulations of weather in the hope of predicting it for months ahead. And it is really rather startling when it comes to dice, even though humanity perversely uses dice as one of its favorite symbols for chance. Dice are just cubes, and a tumbling cube should be no less predictable than an orbiting planet: after all, both objects obey the same laws of mechanical motion. They're different shapes, but equally regular and mathematical ones.

To see how unpredictability can be reconciled with determinism, think about a much less ambitious system than the entire universe—namely, drops of water dripping from a tap.* This is a deterministic system: the principle, the flow of water into the apparatus is steady and uniform, and what happens to it when it emerges is totally prescribed by the laws of fluid motion. Yet a simple but effective experiment demonstrates that this evidently deterministic system can be made to behave unpredictably; and this leads us to some mathematical "lateral thinking," which explains why such a paradox is possible.

If you turn on a tap very gently and wait a few seconds for the flow to settle down, you can usually produce a regular series of drops of water, falling at equally spaced times in a regular rhythm. It would be hard to find anything more predictable than this. But if you slowly turn the tap to

* In the United States: a faucet.

increase the flow, you can set it so that the sequence of drops falls in a very irregular manner, one that sounds random. It may take a little experimentation to succeed, and it helps if the tap turns smoothly. Don't turn it so far that the water falls in an unbroken stream; what you want is a medium-fast trickle. If you get it set just right, you can listen for many minutes without any obvious pattern becoming apparent.

In 1978, a bunch of iconoclastic young graduate students at the University of California at Santa Cruz formed the Dynamical Systems Collective. When they began thinking about this water-drop system, they realized that it's not as random as it appears to be. They recorded the dripping noises with a mircophone and analyzed the sequence of intervals between each drop and the next. What they found was short-term predictability. If I tell you the timing of three successive drops, then you can predict when the next drop will fall. For example, if the last three intervals between drops have been 0.63 seconds, 1.17 seconds, and 0.44 seconds, then you can be sure that the next drop will fall after a further 0.82 seconds. (These numbers are for illustrative purposes only.) In fact, if you know the timing of the first three drops *exactly*, then you can predict the entire future of the system.

So why is Laplace wrong? The point is that we can never measure the initial state of a system exactly. The most precise measurements yet made in any physical system are correct to about ten or twelve decimal places. But Laplace's statement is correct only if we can make measurements to infinite precision, infinitely many decimal places—and of course there's no way to do that. People knew about this problem of measurement error in Laplace's day, but they generally assumed that provided you made the initial measurements to, say ten decimal places, then all subsequent prediction would also be accurate to ten decimal places. The error would not disappear, but neither would it grow.

Unfortunately, it does grow, and this prevents us from stringing together a series of short-term predictions to get one that is valid in the long term. For example, suppose I know the timing of the first three water drops to an accuracy of ten decimal places. Then I can predict the timing of the next drop to nine decimal places, the drop after that to eight decimal places, and so on. At each step, the error grows by a factor of about ten, so I lose confidence in one further decimal place. Therefore, ten steps into the future, I really have no idea at all what the timing of the next drop will be. (Again, the precise figures will probably be different: it may take half a dozen drops to lose one decimal place in accuracy, but even then it takes only sixty drops until the same problem arises.)

This amplification of error is the logical crack through which Laplace's perfect determinism disappears. Nothing short of total perfection of measurement will do. If we could measure the timing to a hundred decimal places, our predictions would fail a mere hundred drops into the future (or six hundred, using the more optimistic estimate). This phenomenon is called "sensitivity to initial conditions," or more informally "the butterfly effect." (When a butterfly in Tokyo flaps its wings, the result may be a hurricane in Florida a month later.) It is intimately associated with a high degree of irregularity of behavior. Anything truly regular is by definition fairly predictable. But sensitivity to initial conditions renders behavior unpredictable—hence irregular. For this reason, a system that displays sensitivity to initial conditions is said to be *chaotic*. Chaotic behavior obeys deterministic laws, but it is so irregular that to the untrained eye it looks pretty much random. Chaos is *not* just complicated, patternless behavior; it is far more subtle. Chaos is *apparently* complicated, *apparently* patternless behavior that actually has a simple, deterministic explanation.

The discovery of chaos was made by many people, too numerous to list here. It came about because of the conjunction of three separate developments. One was a change of scientific focus, away from simple patterns such as repetitive cycles, toward more complex kinds of behavior. The second was the computer, which made it possible to find approximate solutions to dynamical equations easily and rapidly. The third was a new mathematical viewpoint on dynamics—a geometric rather than a numerical viewpoint. The first provided motivation, the second provided technique, and the third provided understanding.

The geometrization of dynamics began about a hundred years ago, when the French mathematician Henri Poincaré—a maverick if ever there was one, but one so brilliant that his views became orthodoxies almost overnight—invented the concept of a phase space. This is an imaginary mathematical space that represents all possible motions of a given dynamical system. To pick a nonmechanical example, consider the population dynamics of a predator-prey ecological system.

The predators are pigs and the prey are those exotically pungent fungi, truffles. The variables upon which we focus attention are the sizes of the two populations—the number of pigs (relative to some reference value such as one million) and the number of truffles (ditto). This choice effectively makes the variables continuous—that is, they take real-number values with decimal places, not just whole-number values. For example, if the reference number of pigs is one million, then a population of 17,439 pigs corresponds to the value 0.017439. Now, the natural growth of truffles depends on how many truffles there are *and* the rate at which pigs eat them: the growth of the pig population depends on how many pigs there are *and* how many truffles they eat. So the rate of change of each variable depends on both variables, an observation that can be turned into a system of differential equations for the population dynamics. I won't write them down, because it's not the equations that matter here: it's what you do with them.

These questions determine—in principle—how any initial population values will change over time. For example, if we start with 17,439 pigs and 788,444 truffles, then you plug in the initial values 0.017439 for the pig variable and 0.788444 for the truffle variable, and the equations implicitly tell you how those numbers will change. The difficulty is to make the implicit become explicit: to *solve* the equations. But in what sense? The natural reflex of a classical mathematician would be to look for a *formula* telling us exactly what the pig population and the truffle population will be at any instant. Unfortunately, such "explicit solutions" are so rare that it is scarcely worth the effort of looking for them unless the equations have a very special and limited form. An alternative is to find approximate solutions on a computer; but that tells us only what will happen for those particular intial values, and most often we want to know what will happen for a lot of different initial values.

Poincaré's idea is to draw a picture that shows what happens for *all* initial values. The state of the system—the sizes of the two populations at some instant of time—can be represented as a point in the plane, using the old trick of coordinates. For example, we might represent the pig population by the horizontal coordinate and the truffle population by the vertical one. The initial state described above corresponds to the point with horizontal coordinate 0.017439 and vertical coordinate 0.788444. Now let time flow. The two coordinates change from one instant to the next, according to the rule expressed by the differential equation, so the corresponding point *moves*. A moving point traces out a curve; and that curve is a visual representation of the future behavior of the entire system. In fact, by looking at the curve, you can "see" important features of the dynamics without worrying about the actual numerical values of the coordinates.

For example, if the curve closes up into a loop, then the two population are following a periodic cycle, repeating the same values over and over again—just as a car on a racetrack keeps going past the same spectator every lap. If the curve homes in toward some particular point and stops, then the populations settle down to a steady state, in which neither changes—like a car that runs out of fuel. By a fortunate coincidence, cycles and steady states are of considerable ecological significance—in particular, they set both upper and lower limits to populations sizes. So the features that the eye detects most easily are precisely the ones that really matter. Moreover, a lot of irrelevant detail can be ignored: for example, we can see that there is a closed loop without having to work out its precise shape (which represents the combined "waveforms" of the two population cycles).

What happens if we try a different pair of initial values? We get a second curve. Each pair of initial values defines a new curve; and we can capture all possible behaviors of the system, for all initial values, by drawing a complete set of such curves. This set of curves resembles the flow lines of an imaginary mathematical fluid, swirling around in the plane. We call the plane the *phase space* of the system, and the set of swirling curves is the system's *phase portrait*. Instead of the symbol-based idea of a differential equation with various initial conditions, we have a geometric, visual scheme of points flowing through pig/truffle space. This differs from an ordinary plane only in that many of its points are potential rather than actual; their coordinates correspond to numbers of pigs and truffles that *could* occur under appropriate initial conditions, but may not occur in a particular case. So as well as the mental shift from symbols to geometry, there is a philosophical shift from the actual to the potential.

The same kind of geometric picture can be imagined for any dynamical system. There is a phase space, whose coordinates are the values of all the variables; and there is a phase portrait, a system of swirling curves that represents all possible behaviors starting from all possible initial conditions, and that are prescribed by the differential equations. This idea constitutes a major advance, because instead of worrying about the precise numerical details of solutions to the equations, we can focus upon the broad sweep of the phase portrait, and bring humanity's greatest asset, its amazing image-processing abilities, to bear. The image of a phase space as a way of organizing the total range of potential behaviors, from among which nature selects the behavior actually observed, has become very widespread in science.

The upshot of Poincaré's great innovation is that dynamics can be visualized in terms of geometric shapes called *attractors*. If you start a dynamical system from some initial point and watch what it does in the long run, you often find that it ends up wandering around on some well-defined shape in phase space. For example, the curve may spiral in toward a closed loop and then go around and around the loop forever. Moreover, different choices of initial conditions may lead to the same final shape. If so, that shape is known as an attractor. The long-term dynamics of a system is governed by its attractors, and the shape of the attractor determines what type of dynamics occurs.

For example, a system that settles down to a steady state has an attractor that is just a point. A system that settles down to repeating the same behavior periodically has an attractor that is a closed loop. That is, closed loop attractors correspond to oscillators. Recall the description of a vibrating violin string from chapter 5; the string undergoes a sequence of motions that eventually puts it back where it started, ready to repeat the sequence over and over forever. I'm not suggesting that the violin string moves in a physical loop. But my *description* of it is a closed loop in a metaphorical sense: the motion takes a round trip through the dynamic landscape of phase space.

Chaos has its own rather weird geometry: it is associated with curious fractal shapes called *strange attractors*. The butterfly effect implies that the detailed motion *on* a strange attractor can't be determined in advance. But this doesn't alter the fact that it *is* an attractor. Think of releasing a Ping-Pong ball into a stormy sea. Whether you drop it from the air or release it from underwater, it moves toward the surface. Once on the surface, it follows a very complicated path in the surging waves, but however complex that path is, the ball stays on—or at least very near—the surface. In this image, the surface of the sea is an attractor. So, chaos notwithstanding, no matter what the starting point may be, the system will end up very close to its attractor.

Chaos is well established as a mathematical phenomenon, but how can we detect it in the real world? We must perform experiments—and there is a problem. The traditional role of experiments in science is to test theoretical predictions, but if the butterfly effect is in operation—as it is for any chaotic system—how can we hope to test a prediction? Isn't chaos inherently untestable, and therefore unscientific?

The answer is a resounding no, because the word "prediction" has two meanings. One is "foretelling the future," and the butterfly effect prevents this when chaos is present. But the other is "describing in advance what the outcome of an experiment will be." Think about tossing a coin a hundred times. In order to predict—in the fortune-teller's sense—what happens, you must list in advance the result of each of the tosses. But you can make scientific predictions, such as "roughly half the coins will show heads," *without* foretelling the future in detail—even when, as here, the system is random. Nobody suggests that statistics is unscientific because it deals with unpredictable events, and therefore chaos should be treated in the same manner. You can make all sorts of predictions about a chaotic system; in fact, you can make enough predictions to distinguish deterministic chaos from true randomness. One thing that you can often predict is the shape of the attractor, which is not altered by the butterfly effect. All the butterfly effect does is to make the system follow different paths on the same attractor. In consequence, the general shape of the attractor can often be inferred from experimental observations.

The discovery of chaos has revealed a fundamental misunderstanding in our views of the relation between rules and the behavior they produce—between cause and effect. We used to think that deterministic causes must produce regular effects, but now we see that they can produce highly irregular effects that can easily be mistaken for randomness. We used to think that simple causes must produce simple effects (implying that complex effects must have complex causes), but now we know that simple causes can produce complex effects. We realize that knowing the rules is not the same as being able to predict future behavior.

How does this discrepancy between cause and effect arise? Why do the same rules sometimes produce obvious patterns and sometimes produce chaos? The answer is to be found in every kitchen, in the employment of that simple mechanical device, an eggbeater. The motion of the two beaters is simple and predictable, just as Laplace would have expected: each beater rotates steadily. The motion of the sugar and the egg white in the bowl, however, is far more complex. The two ingredients get mixed up—that's what eggbeaters are for. But the two rotary beaters *don't* get mixed up—you don't have to disentangle them from each other when you've finished. Why is the motion of the incipient meringue so different from that of the beaters? Mixing is a far more complicated, dynamic process than we tend to think. Imagine trying to predict where a particular grain of sugar will end up! As the mixture passes between the pair of beaters, it is pulled apart, to left and right, and two sugar grains that start very close together soon get a long way apart and follow independent paths. This is, in fact, the butterfly effect in action—tiny changes in initial conditions have big effects. So mixing is a chaotic process.

Conversely, every chaotic process involves a kind of mathematical mixing in Poincaré's imaginary phase space. This is why tides are predictable but weather is not. Both involve the same kind of mathematics, but the dynamics of tides does not get phase space mixed up, whereas that of the weather does.

It's not what you do, it's the way that you do it.

Chaos is overturning our comfortable assumption about how the world works. It tells us that the universe is far stranger than we think. It casts doubt on many traditional methods of science: merely knowing the laws of nature is no longer enough. On the other hand, it tells us that some things that we thought were just random may actually be consequences of simple laws. Nature's chaos is bound by rules. In the past, science tended to ignore events or phenomena that seemed random, on the grounds that since they had no obvious patterns they could not be governed by simple laws. Not so. There are simple laws right under our noses—laws governing disease epidemics, or heart attacks, or plagues of locusts. If we learn those laws, we may be able to prevent the disasters that follow in their wake.

Already chaos has shown us new laws, even new types of laws. Chaos contains its own brand of new universal patterns. One of the first to be discovered occurs in the dripping tap. Remember that a tap can drip rhythmically or chaotically, depending on the speed of the flow. Actually, both the regularly dripping tap and the "random" one are following slightly different variants of the same mathematical prescription. But as the rate at which water passes through the tap increases, the type of dynamics changes. The attractor in phase space that represents the dynamics keeps changing—and it changes in a predictable but highly complex manner.

Start with a regularly dripping tap: a repetitive drip-drip-drip-drip rhythm, each drop just like the previous one. Then turn the tap slightly, so that the drips come slightly faster. Now the rhythm goes drip-DRIP-drip-DRIP, and repeats every two drops. Not only the size of the drop, which governs how loud the drip sounds, but also the timing changes slightly from one drop to the next.

If you allow the water to flow slightly faster still, you get a four-drop rhythm: drip-DRIP-*drip*-DRIP. A little faster still, and you produce an eight-drop rhythm: drip-DRIP-*drip*-*DRIP*-**drip-DRIP-drip-DRIP.**The length of the repetitive sequence of drops keeps on doubling. In a mathematical model, this process continues indefinitely, with rhythmic groups of 16,32,64 drops, and so on. But it takes tinier and tinier changes to the flow rate to produce each successive doubling of the period; and there is a flow rate by which the size of the group has doubled infinitely often. At this point, no sequence of drops repeats exactly the same pattern. This is chaos.

We can express what is happening in Poincaré's geometric language. The attractor for the tap begins as a closed loop, representing a periodic cycle. Think of the loop as an elastic band wrapped around your finger. As the flow rate increases, this loop splits into two nearby loops, like an elastic band wound twice around your finger. This band is twice as long as the original, which is why the period is twice as long. Then in exactly the same way, this already-doubled loop doubles again, all the way along its length, to create the period-four cycle, and so on. After infinitely many doublings, your finger is decorated with elastic spaghetti, a chaotic attractor.

This scenario for the creation of chaos is called a period-doubling cascade. In 1975, the physicist Mitchell Feigenbaum discovered that a particular number, which can be measured in experiments, is associated with every period-doubling cascade. The number is roughly 4.669, and it ranks alongside π (pi) as one of those curious numbers that seem to have extraordinary significance in both mathematics and its relation to the natural world. Feigenbaum's number has a symbol, too: the Greek letter δ (delta). The number π tells us how the circumference of a circle relates to its diameter. Analogously, Feigenbaum's number δ tells us how the period of the drips relates to the rate of flow of the water. To be precise, the extra amount by which you need to turn on the tap decreases by a factor of 4.669 at each doubling of the period.

The number π is a quantitative signature for anything

involving circles. In the same way, the Feigenbaum number δ is a quantitative signature for any period-doubling cascade, no matter how it is produced or how it is realized experimentally. That very same number shows up in experiments on liquid helium, water, electronic circuits, pendulums, magnets, and vibrating train wheels. It is a new universal pattern in nature, one that we can see only through the eyes of chaos; a *quantitative* pattern, a number, emerges from a qualitative phenomenon. One of nature's numbers, indeed. The Feigenbaum number has opened the door to a new mathematical world, one we have only just begun to explore.

The precise pattern found by Feigenbaum, and other patterns like it, is a matter of fine detail. The basic point is that even when the consequence of natural laws seem to be patternless, the laws are still there and so are the patterns. Chaos is not random: it is *apparently* random behavior resulting from precise rules. Chaos is a cryptic form of order.

Science has traditionally valued order, but we are beginning to appreciate the fact that chaos can offer science distinct advantages. Chaos makes it much easier to respond quickly to an outside stimulus. Think of tennis players waiting to receive a serve. Do they stand still? Do they move regularly from side to side? Of course not. They dance erratically from one foot to the other. In part, they are trying to confuse their opponents, but they are also getting ready to respond to any serve sent their way. In order to be able to move quickly in any particular direction, they make rapid movements in many different directions. A chaotic system can react to outside events much more quickly, and with much less effort, than a nonchaotic one. This is important for engineering control problems. For example, we now know that some kinds of turbulence result from chaos—that's what makes turbulence look random. It may prove possible to make the airflow past an aircraft's skin much less turbulent, and hence less resistant to motion, by setting up control mechanisms that respond extremely rapidly to cancel out any small regions of incipient turbulence. Living creatures, too, must behave chaotically in order to respond rapidly to a changing environment.

This idea has been turned into an extremely useful practical technique by a group of mathematicians and physicists, among them William Ditto, Alan Garfinkel, and Jim Yorke: they call it chaotic control. Basically, the idea is to make the butterfly effect work for you. The fact that small changes in initial conditions create large changes in subsequent behavior can be an advantage; all you have to do is ensure that you get the large changes you want. Our understanding of how chaotic dynamics works makes it possible to devise control strategies that do precisely this. The methods has had several successes. Space satellites use a fuel called hydrazine to make course corrections. One of the earliest successes of chaotic control was to divert a dead satellite from its orbit and send it out for an encounter with an asteroid, using only the tiny amount of hydrazine left on board. NASA arranged for the satellite to swing around the Moon five times, nudging it slightly each time with a tiny shot of hydrazine. Several such encounters were achieved, in an operation that successfully exploited the occurrence of chaos in the three-body problem (here, Earth/Moon/satellite) and the associated butterfly effect.

The same mathematical idea has been used to control a magnetic ribbon in a turbulent fluid—a prototype for controling turbulent flow past a submarine or an aircraft. Chaotic control has been used to make erratically beating hearts return to a regular rhythm, presaging invention of the intelligent pacemaker. Very recently, it has been used both to set up and to prevent rhythmic waves of electrical activity in brain tissue, opening up the possibility of preventing epileptic attacks.

Chaos is a growth industry. Every week sees new discoveries about the underlying mathematics of chaos, new applications of chaos to our understanding of the natural world, or new technological uses of chaos—including the chaotic dishwasher, a Japanese invention that uses two rotating arms, spinning chaotically, to get dishes cleaner using less energy; and a British machine that uses chaos-theoretic data analysis to improve quality control in spring manufacture.

Much, however, remains to be done. Perhaps the ultimate unsolved problem of chaos is the strange world of the quantum, where Lady Luck rules. Radioactive atoms decay "at random"; their only regularities are statistical. A large quantity of radioactive atoms has a well-defined half-life—a period of time during which half the atoms will decay. But we can't predict which half. Albert Einstein's protest, mentioned earlier, was aimed at just this question. Is there really *no* difference at all between a radioactive atom that is not going to decay, and one that's just about to? Then *how does the atom know what to do*?

Might the apparent randomness of quantum mechanics be fraudulent? Is it really deterministic chaos? Think of an atom as some kind of vibrating droplet of cosmic fluid. Radioactive atoms vibrate very energetically, and every so often a smaller drop can split off—decay. The vibrations are so rapid that we can't measure them in detail: we can only measure averaged quantities, such as energy levels. Now, classical mechanics tells us that a drop of real fluid can vibrate chaotically. When it does so, its motion is deterministic but unpredictable. Occasionally, "at random," the vibrations conspire to split off a tiny droplet. The butterfly effect makes it impossible to say in advance just when the drop will split; but that event has precise statistical features, including a well defined half-life.

Could the apparently random decay of radioactive atoms be something similar, but on a microcosmic scale? After all, why are there any statistical regularities at all? Are they traces of an underlying determinism? Where *else* can statistical regularities come from? Unfortunately, nobody has yet made this seductive idea work—though it's similar in spirit to the fashionable theory of superstrings, in which a subatomic particle is a kind of hyped-up vibrating multidimensional loop. The main similar feature here is that both the vibrating loop and the vibrating drop introduce new "internal variables" into the physical picture. A significant difference is the way these two approaches handle quantum indeterminacy. Superstring theory, like conventional quantum mechanics, sees this determinacy as being genuinely random. In a system like the drop, however, the apparent indeterminacy is actually generated by a deterministic, but chaotic, dynamic. The trick—if only we knew how to do it—would be to invent some kind of structure that retains the successful feature of superstring theory, while making some of the internal variables behave chaotically. It would be an appealing way to render the Deity's dice deterministic, and keep the shade of Einstein happy.

CHAPTER 9

DROPS, DYNAMICS, AND DAISIES

Chaos teaches us that systems obeying simple rules can behave in surprisingly complicated ways. There are important lessons here for everybody—managers who imagine that tightly controlled companies will automatically run smoothly, politicians who think that legislating against a problem will automatically eliminate it, and scientists who imagine that once they have modeled a system their work is complete. But the world cannot be totally chaotic, otherwise we would not be able to survive in it. In fact, one of the reasons that chaos was not discovered sooner is that in many ways our world is simple. That simplicity tends to disappear when we look below the surface, but on the surface it is still there. Our use of language to describe our world rests upon the existence of underlying simplicities. For example, the statement "foxes chase rabbits" makes sense only because it captures a general pattern of animal interaction. Foxes *do* chase rabbits, in the sense that if a hungry fox sees a rabbit then it is likely to run after it.

However, if you start to look at the details, they rapidly become so complicated that the simplicity is lost. For example, in order to perform this simple act, the fox must recognize the rabbit as a rabbit. Then it must put its legs into gear to run after it. In order to comprehend these actions, we must understand vision, pattern recognition in the brain, and locomotion. In chapter 7, we investigated the third item, locomotion, and there we found the intricacies of physiology and neurology—bones, muscles, nerves, and brains. The action of muscles in turn depends on cell biology and chemistry; chemistry depends on quantum mechanics; and quantum mechanics may, in turn, depend on the much-sought Theory of Everything, in which all of the laws of physics come together in a single unified whole. If instead of locomotion we pursue the path opened by vision or pattern recognition, we again see the same kind of ever-branching complexity.

The task looks hopeless—except that the simplicities we start from exist, so either nature uses this enormously complex network of cause and effect or it sets things up so that most of the complexity doesn't matter. Until recently, the natural paths of investigation in science led deeper and deeper into the tree of complexity—what Jack Cohen and I have called the "reductionist nightmare."* We have learned a lot about nature by going that route—especially regarding how to manipulate it to our own ends. But we have lost sight of the big simplicities because we no longer see them as being simple at all. Recently, a radically

* In *The Collapse of Chaos.*

different approach has been advocated, under the name *complexity theory*. Its central theme is that large-scale simplicities emerge from the complex interactions of large numbers of components.

In this final chapter, I want to show you three examples of simplicity emerging from complexity. They are not taken from the writings of the complexity theorists; instead I have chosen them from the mainstream of modern applied mathematics, the theory of dynamical systems. There are two reasons why I have done this. One is that I want to show that the central philosophy of complexity theory is popping up all over science, independently of any explicit movement to promote it. There is a quiet revolution simmering, and you can tell because the bubbles are starting to break the surface. The other is that each piece of work solves a long-standing puzzle about mathematical patterns in the natural world—and in so doing opens our eyes to features of nature that we would not otherwise have appreciated. The three topics are the shape of water drops, the dynamic behavior of animal populations, and the strange patterns in plant-petal numerology, whose solution I promised in the opening chapter.

To begin, let us return to the question of water dripping slowly from a tap. Such a simple, everyday phenomenon—yet it has already taught us about chaos. Now it will teach us something about complexity. This time we do not focus on the timing of successive drops. Instead, we look at what shape the drop takes up as it detaches from the end of the tap.

Well, it's obvious, isn't it? It must be the classic "teardrop" shape, rather like a tadpole; round at the head and curving away to a sharp tail. After all, that's why we call such a shape a teardrop.

But it's not obvious. In fact, it's not true.

When I was first told of this problem, my main surprise was that the answer had not been found long ago. Literally miles of library shelves are filled with scientific studies of fluid flow; surely somebody took the trouble to look at the shape of a drop of water? Yet the early literature contains only one correct drawing, made over a century ago by the physicist Lord Rayleigh, and is so tiny that hardly anybody noticed it. In 1990, the mathematician Howell Peregrine and colleagues at Bristol University photographed the process and discovered that it is far more complicated—but also far more interesting—than anybody would ever imagine.

The formation of the detached drop begins with a bulging droplet hanging from a surface, the end of the tap. It develops a waist, which narrows, and the lower part of the droplet appears to be heading toward the classic teardrop shape. But instead of pinching off to form a short, sharp tail, the waist lengthens into a long thin cylindrical thread with an almost spherical drop hanging from its end. Then the thread starts to narrow, right at the point where it meets the sphere, until it develops a sharp point. At this stage, the general shape is like a knitting needle that is just touching an orange. Then the orange falls away from the needle, pulsating slightly as it falls. But that's only half the story. Now the sharp end of the needle begins to round off, and tiny waves travel back up the needle toward its root, making it look like a string of pearls that become tinier and tinier. Finally, the hanging thread of water narrows to a sharp point at the top end, and it, too, detaches. As it falls, its top end rounds off and a complicated series of waves travels along it.

I hope you find this as astonishing as I do. I had never imagined that falling drops of water could be so *busy*.

These observations make it clear why nobody had previously studied the problem in any great mathematical detail. It's too hard. When the drop detaches, there is a singularity in the problem—a place where the mathematics becomes very nasty. The singularity is the tip of the "needle." But why is there a singularity at all? Why does the drop detach in such a complex manner? In 1994, J. Eggers and T. F. Dupont showed that the scenario is a consequence of the equations of fluid motion. They simulated those equations on a computer and reproduced Peregrine's scenario.

It was a brilliant piece of work. But in some respects it does not provide a complete answer to my question. It is reassuring to learn that the equations of fluid flow do

FIGURE 4.
The shapes taken by a falling drop of water as it becomes detached.

predict the correct scenario, but that in itself doesn't help me understand why that scenario happens. There is a big difference between calculating nature's numbers and getting your brain around the answer—as Majikthise and Vroomfondel discovered when the answer was "Forty-two."

Further insight into the mechanism of detaching drops

has come about through the work of X. D. Shi, Michael Brenner, and Sidney Nagel, of the University of Chicago. The main character in the story was already present in Peregrine's work: it is a particular kind of solution to the equations of fluid flow called a "similarity solution." Such a solution has a certain kind of symmetry that makes it mathematically tractable: it repeats its structure on a smaller scale after a short interval of time. Shi's group took this idea further, asking how the shape of the detaching drop depends on the fluid's viscosity. They performed experiments using mixtures of water and glycerol to get different viscosities. They also carried out computer simulations and developed the theoretical approach via similarity solutions. What they discovered is that for more viscous fluids, a second narrowing of the thread occurs before the singularity forms and the drop detaches. You get something more like an orange suspended by a length of string from the tip of a knitting needle. At higher viscosities still, there is a third narrowing—an orange suspended by a length of cotton from a length of string from the tip of a knitting needle. And as the viscosity goes up, so the number of successive narrowings increases without limit—at least, if we ignore the limit imposed by the atomic structure of matter.

Amazing!

The second example is about population dynamics. The use of that phrase reflects a long tradition of mathematical modeling in which the changes in population of interacting creatures are represented by differential equations. My pig/truffle system was an example. However, there is a lack of biological realism in such models—and not just as regards my choice of creatures. In the real world, the mechanism that governs population sizes is not a "law of population," akin to Newton's law of motion. There are all kinds of other effects: for example, random ones (can the pig dig out the truffle or is there a rock in the way?) or types of variability not included in the equations (some pigs habitually produce more piglets than others).

In 1994, Jacquie McGlade, David Rand, and Howard Wilson, of Warwick University, carried out a fascinating study that bears on the relation between more biologically realistic models and the tradional equations. It follows a strategy common in complexity theory: set up a computer simulation in which large numbers of "agents" interact according to biologically plausible (though much simplified) rules, and try to extract large-scale patterns from the results of that simulation. In this case, the simulation was carried out by means of a "cellular automaton," which you can think of as a kind of mathematical computer game. McGlade, Rand, and Wilson, lacking my bias in favor of pigs, considered the more traditional foxes and rabbits. The computer screen is divided into a grid of squares, and each square is assigned a color—say, red for a fox, gray for a rabbit, green for grass, black for bare rock. Then a system of rules is set up to model the main biological influences at work. Examples of such rules might be:

- If a rabbit is next to grass, it moves to the position of the grass and eats it.
- If the fox is next to a rabbit, it moves to the position of the rabbit and eats it.
- At each stage of the game, a rabbit breeds new rabbits with some chosen probability.
- A fox that has not eaten for a certain number of moves will die.

McGlade's group played a more complicated game than this, but you get the idea. Each move in the game takes the current configuration of rabbits, foxes, grass, and rock, and applies the rules to generate the next configuration—tossing computer "dice" when random choices are required. The process continues for several thousand moves, an "artificial ecology" that plays out the game of life on a computer screen. This artificial ecology resembles a dynamical system, in that it repeatedly applies the same bunch of rules; but it also includes random effects, which places the model in a different mathematical category altogether: that of stochastic cellular automata—computer games with chance.

Precisely because the ecology is an artificial one, you can perform experiments that are impossible, or too expensive, to perform in real ecology. You can, for example, watch how the rabbit population in a given region changes over time, and get the exact numbers. This is where McGlade's group made a dramatic and surprising discovery. They realized that if you look at too tiny a region, what you see is largely random. For example, what happens on a single square looks extremely complicated. On the other hand, if you look at too large a region, all you see is the statistics of the population, averaged out. On intermediate scales, though, you may see something less dull. So they developed a technique for finding the size of region that would provide the largest amount of interesting information. They then observed a region of that size and recorded the changing rabbit population. Using methods developed in chaos theory, they asked whether that series of numbers was deterministic or random, and if deterministic, what its attractor looked like. This may seem a strange thing to do, inasmuch as we know that the rules for the simulation build in a great deal of randomness, but they did it anyway.

What they found was startling. Some 94 percent of the dynamics of the rabbit population on this intermediate scale can be accounted for by deterministic motion on a chaotic attractor in a four-dimensional phase space. In short, a differential equation with only four variables captures the important features of the dynamics of the rabbit population with only a 6-percent error—despite the far greater complexities of the computer-game model. This discovery implies that models with small numbers of variables may be more "realistic" than many biologists have hitherto assumed. Its deeper implication is that simple large-scale features can and do emerge from the fine structure of complex ecological games.

My third and final example of a mathematical regularity of nature that emerges from complexity rather than having been "built in with the rules" is the number of petals of flowers. I mentioned in chapter 1 that the majority of plants have a number of petals taken from the series 3,5,8,13,21,34,55,89. The view of conventional biologists is that the flower's genes specify all such information, and that's really all there is to it. However, just because living organisms have complicated DNA sequences that determine which proteins they are made of, and so on, it doesn't follow that genes determine everything. And even if they do, they may do so only indirectly. For example, genes tell plants how to make chlorophyll, but they don't tell the plants what color the chlorophyll has to be. If it's chlorophyll, it's green—there's no choice. So some features of the morphology of living creatures are genetic in origin and some are a consequence of physics, chemistry, and the dynamics of growth. One way to tell the difference is that genetic influences have enormous flexibility, but physics, chemistry, and dynamics produce mathematical regularities.

The numbers that arise in plants—not just for petals but for all sorts of other features—display mathematical regularities. They form the beginning of the so-called Fibonacci series, in which each number is the sum of the two that precede it. Petals aren't the only places you find Fibonacci numbers, either. If you look at a giant sunflower, you find a remarkable pattern of florets—tiny flowers that eventually become seeds—in its head. The florets are arranged in two intersecting families of spirals, one winding clockwise, the other counter-clockwise. In some species the number of clockwise spirals is thirty-four, and the number of counterclockwise spirals is fifty-five. Both are Fibonacci numbers, occurring consecutively in the series. The precise numbers depend on the species of sunflower, but you often get 34 and 55, or 55 and 89, or even 89 and 144, the next Fibonacci number still. Pineapples have eight rows of scales—the diamond-shaped markings—sloping to the left, and thirteen sloping to the right.

Leonardo Fibonacci, in about 1200, invented his series in a problem about the growth of a population of rabbits. It wasn't as realistic a model of rabbit-population dynamics as the "game of life" model I've just discussed, but it was a very interesting piece of mathematics nevertheless, because it was the first model of its kind and because mathematicians find Fibonacci numbers fascinating and beautiful in their own right. The key question for this chapter is this: If genetics can choose to give a flower any number of petals it likes, or a pine cone any number of scales that it likes, why do we observe such a preponderance of Fibonacci numbers?

The answer, presumably, has to be that the numbers arise through some mechanism that is more mathematical than arbitrary genetic instructions. The most likely candidate is some kind of dynamic constraint on plant development, which naturally leads to Fibonacci numbers. Of course, appearances may be deceptive, it *could* be all in the genes. But if so, I'd like to know how the Fibonacci numbers got turned into DNA codes, and why it was those numbers. Maybe evolution started with the mathematical patterns that occurred naturally, and fine-tuned them by natural selection. I suspect a lot of that has happened—tigers' stripes, butterflies' wings. That would explain why geneticists are convinced the patterns are genetic and mathematicians keep insisting they are mathematical.

The arrangement of leaves, petals, and the like in plants has a huge and distinguished literature. But early approaches are purely descriptive—they don't explain how the numbers relate to plant growth, they just sort out the geometry of the arrangements. The most dramatic insight yet comes from some very recent work of the French mathematical physicists Stéphane Douady and Yves Couder. They devised a theory of the dynamics of plant growth and used computer models and laboratory experiments to show that it accounts for the Fibonacci pattern.

The basic idea is an old one. If you look at the tip of the shoot of a growing plant, you can detect the bits and pieces from which all the main features of the plant—leaves, petals, sepals, florets, or whatever—develop. At the center of the tip is a circular region of tissue with no special features, called the apex. Around the apex, one by one, tiny lumps form, called primordia. Each primordium migrates away from the apex—

or, more accurately, the apex grows away from the lump—and eventually the lump develops into a leaf, petal, or the like. Moreover, the general arrangement of those features is laid down right at the start, as the primordia form. So basically all you have to do is explain why you see spiral shapes and Fibonacci numbers in the primordia.

The first step is to realize that the spirals most apparent to the eye are not fundamental. The most important spiral is formed by considering the primordia in their order of appearance. Primordia that appear earlier migrate farther, so you can deduce the order of appearance from the distance away from the apex. What you find is that successive primordia are spaced rather sparsely along a tightly wound spiral, called the generative spiral. The human eye picks out the Fibonacci spirals because they are formed from primordia that appear near each other in space; but it is the sequence in time that really matters.

The essential quantitative feature is the angle between successive primordia. Imagine drawing lines from the centers of successive primordia to the center of the apex and measuring the angle between them. Successive angles are pretty much equal; their common value is called the divergence angle. In other words, the primordia are equally spaced—in an angular sense—along the generative spiral. Moreover, the divergence angle is usually very close to 137.5°, a fact first emphasized in 1837 by the crystallographer Auguste Bravais and his brother Louis. To see why that number is significant, take two consecutive numbers in the Fibonacci series: for example, 34 and 55. Now form the corresponding fraction 34/55 and multiply by 360°, to get 222.5°. Since this is more than 180°, we should measure it in the opposite direction round the circle—or, equivalently, subtract it from 360°. The result is 137.5°, the value observed by the Bravais brothers.

The ratio of consecutive Fibonacci numbers gets closer and closer to the number 0.618034. For instance, 34/55 = 0.6182 which is already quite close. The limiting value is exactly $(\sqrt{5}-1)/2$, the so-called golden number, often denoted by the Greek letter phi (ϕ). Nature has left a clue for mathematical detectives: the angle between successive primordia is the "golden angle" of $360(1-\phi)° = 137.5°$. In 1907, G. Van Iterson followed up this clue and worked out what happens when you plot successive points on a tightly wound spiral separated by angles of 137.5°. Because of the way neighboring points align, the human eye picks out two families of interpenetrating spirals—one winding clockwise and the other counterclockwise. And because of the relation between Fibonacci numbers and the golden number, the numbers of spirals in the two families are consecutive Fibonacci numbers. *Which* Fibonacci numbers depends on the tightness of the spiral. How does that explain the numbers of petals? Basically, you get one petal at the outer edge of each spiral in just one of the families.

At any rate, it all boils down to explaining why successive

primordia are separated by the golden angle: then everything else follows.

Douady and Couder found a dynamic explanation for the golden angle. They built their ideas upon an important insight of H. Vogel, dating from 1979. His theory is again a descriptive one—it concentrates on the geometry of the arrangement rather than on the dynamics that caused it. He performed numerical experiments which strongly suggested that if successive primordia are placed along the generative spiral using the golden angle, they will pack together most efficiently. For instance, suppose that, instead of the golden angle, you try a divergence angle of 90°, which divides 360° exactly. Then successive primordia are arranged along four radial lines forming a cross. In fact, if you use a divergence angle that is a rational multiple of 360°, you always get a system of radial lines. So there are gaps between the lines and the primordia don't pack efficiently. Conclusion: to fill the space efficiently, you need a divergence angle that is an irrational multiple of 360°—a multiple by a number that is not an exact fraction. But which irrational number? Numbers are either irrational or not but—like equality in George Orwell's *Animal Farm*—some are more irrational than others. Number theories have long known that the most irrational number is the golden number. It is "badly approximable" by rational numbers, and if you quantify how badly, it's the worst of them all. Which, turning the argument on its head, means that a golden divergence angle should pack the primordia most

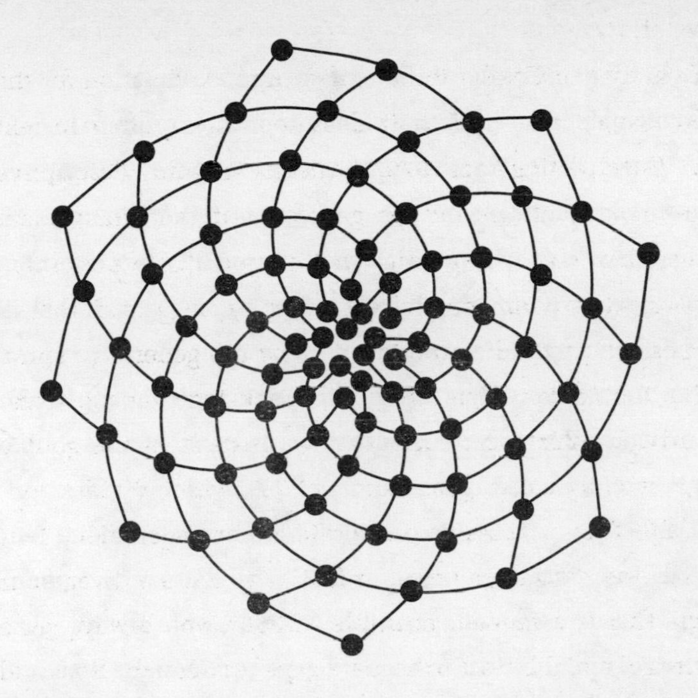

FIGURE 5.
Successive dots arranged at angles of 137.5° to each other along a tightly wound spiral (not shown) naturally fall into two families of loosely wound spirals that are immediately apparent to the eye. Here there are 8 spirals in one direction and 13 in the other—consecutive Fibonacci numbers.

closely. Vogel's computer experiments confirm this expectation but do not prove it with full logical rigor.

The most remarkable thing Douady and Couder did was to obtain the golden angle as a consequence of simple dynamics rather then to postulate it directly on grounds of efficient packing. They assumed that successive elements of some kind—representing primordia—form at equally spaced intervals of time somewhere on the rim of a small circle, representing the apex; and that these elements then migrate radially at some specified initial velocity. In addition, they assume that the elements repel each other—like equal electric charges, or magnets with the same polarity. This ensures that the radial motion keeps going and that each new element appears as far as possible from its immediate predecessors. It's good bet that such a system will satisfy Vogel's criterion of efficient packing, so you would expect the golden angle to show up of its own accord. And it does.

Douady and Couder performed an experiment—not with plants, but using a circular dish full of silicone oil packed in a vertical magnetic field. They let tiny drops of magnetic fluid fall at regular intervals of time into the center of the dish. The drops were polarized by the magnetic field and repelled each other. They were given a boost in the radial direction by making the magnetic field stronger at the edge of the dish than it was in the middle. The patterns that appeared depended on how big the intervals between drops were; but a very prevalent pattern was one in which successive drops lay on a spiral with divergence angle very close to the golden angle, giving a sunflower-seed pattern of interlaced spirals. Douady and Couder also carried out computer calculations, with similar results. By both methods, they found that the divergence angle depends on the interval between drops according to a complicated branching pattern of wiggly curves. Each section of a curve between successive wiggles corresponds to a particular pair of numbers of spirals. The main branch is very close to a divergence angle of 137.5°, and along it you find all possible pairs of consecutive Fibonacci numbers, one after the other in numerical sequence. The gaps between branches represent "bifurcations," where the dynamics undergoes significant changes.

Of course, nobody is suggesting that botany is quite as perfectly mathematical as this model. In particular, in many plants the rate of appearance of primordia can speed up or slow down. In fact, changes in morphology—whether a given primordium becomes a leaf or a petal, say—often accompany such variations. So maybe what the genes do is affect the timing of the appearance of the primordia. But plants don't need their genes to tell them how to space their primordia: that's done by the dynamics. It's a partnership of physics and genetics, and you need both to understand what's happening.

Three examples, from very different parts of science. Each, in its own way, an eye-opener. Each a case study in the origins of nature's numbers—the deep mathematical regularities that can be detected in natural forms. And there is a common thread, an even deeper message, buried within them. Not that nature is complicated. No, nature is, in its own subtle way, simple. However, those simplicities do not present themselves to us directly. Instead, nature leaves clues for the mathematical detectives to puzzle over. It's a fascinating game, even to a spectator. And it's an absolutely irresistible one if you are a mathematical Sherlock Holmes.

EPILOGUE

MORPHOMATICS

I have another dream.

My first dream, the Virtual Unreality Machine, is just a piece of technology. It would help us visualize mathematical abstractions, encourage us to develop new intuition about them, and let us ignore the tedious bookkeeping parts of mathematical inquiry. Mostly, it would make it easier for mathematicians to explore their mental landscape. But because they sometimes create new bits of that landscape as they wander around it, the Virtual Unreality Machine would play a creative role, too. In fact, it—or something like it—will soon exist.

I call my second dream "morphomatics." It is not a matter of technology; it is a way of thinking. Its creative importance would be immense. But I have no idea whether it will ever come into being, or even whether it is possible.

I hope it is, because we need it.

The three examples in the previous chapter—liquid drops, foxes and rabbits, and petals—are very different in detail, but they illustrate the same philosophical point about how the universe works. It does not go *directly* from simple laws, like the laws of motion, to simple patterns, like the elliptical orbits of the planets. Instead, it passes through an enormous tree of ramifying complexity, which somehow collapses out again into relatively simple patterns on appropriate scales. The simple statement "a drop falls off the tap" is accomplished by way of an amazingly complex and surprising sequence of transitions. We do not yet know *why* those transitions derive from the laws of fluid flow, although we have computer evidence that they do. The effect is simple, the cause is not. The foxes, rabbits, and grass play a mathematical computer game with complicated and probabilistic rules. Yet important features of their artificial ecology can be represented to 94-percent accuracy by a dynamical system with four variables. And the number of petals on a plant is a consequence of a complex dynamic interaction between all the primordia, which just happens to lead, via the golden angle, to Fibonacci numbers. The Fibonacci numbers are clues for the mathematical Sherlock Holmeses to follow up—they are not the master criminal behind those clues. In this case, the mathematical Moriarty is dynamics, not Fibonacci—nature's mechanisms, not nature's numbers.

There is a common message in these three mathematical tales: nature's patterns are "emergent phenomena." They emerge from an ocean of complexity like Botticelli's Venus from her half shell—unheralded, transcending their origins. They are not *direct* consequences of the deep simplicities of natural laws; those laws operate on the wrong level for that. They are without doubt *indirect* consequences of the deep simplicities of nature, but the route from cause to effect becomes so complicated that no one can follow every step of it.

If we really want to come to grips with the emergence of pattern, we need a new approach to science, one that can stand alongside the traditional emphasis on the underlying laws and equations. Computer simulations are a part of it, but we need more. It is not satisfying to be told that some pattern occurs because the computer says so. We want to know *why*. And that means we must develop a new kind of mathematics, one that deals with patterns as patterns and not just as accidental consequences of fine-scale interactions.

I don't want us to replace current scientific thinking, which has brought us a long, long way. I want us to develop something that complements it. One of the most striking features of recent mathematics has been its emphasis on general principles and abstract structures—on the qualitative rather than the quantitative. The great physicist Ernest Rutherford once remarked that "qualitative is just poor quantitative," but that attitude no longer makes much sense. To turn Rutherford's dictum on its head, quantitative is just poor qualitative. Number is just one of an enormous variety of mathematical qualities that can help us understand and describe nature. We will never understand the growth of a tree or the dunes in the desert if we try to reduce all of nature's freedom to restrictive numerical schemes.

The time is ripe for the development of a new kind of mathematics, one that possesses the kind of intellectual rigor that was the real point of Rutherford's criticism of sloppy qualitative reasoning, but has far more conceptual flexibility. We need an effective mathematical theory of form, which is why I call my dream "morphomatics." Unfortunately, many branches of science are currently headed in the opposite direction. For example, DNA programming is often held to be the sole answer to form and pattern in organisms. However, current theories of biological development do not adequately explain why the organic and inorganic worlds share so many mathematical patterns. Perhaps DNA encodes dynamic rules for development rather than just encoding the final developed form. If so, our current theories ignore crucial parts of the developmental process.

The idea that mathematics is deeply implicated in natural form goes back to D'Arcy Thompson; indeed, it goes back to the ancient Greeks, maybe even to the Babylonians. Only very recently, however, have we started to develop the right *kind* of mathematics. Our previous mathematical schemes were themselves too inflexible, geared to the constraints of pencil and paper. For example, D'Arcy Thompson noticed similarities between the shapes of various organisms and the flow patterns of fluids, but fluid dynamics as currently understood uses equations that are far too simple to model organisms.

If you watch a single-celled creature under a microscope, the most amazing thing you see is the apparent sense of *purpose* in the way it flows. It really does look as if it knows where it is going. Actually, it is responding in a very specific way to its environment and its own internal state. Biologists are beginning to unravel the mechanisms of cellular motion, and these mechanisms are a lot more complex than classical fluid dynamics. One of the most important features of a cell is the so-called cytoskeleton, a tangled network of tubes that resembles a bale of straw and provides the cell interior with a rigid scaffolding. The cytoskeleton is amazingly flexible and dynamic. It can disappear altogether, under the influence of certain chemicals, or it can be made to grow wherever support is required. The cell moves about by tearing down its cytoskeleton and putting it up somewhere else.

The main component of the cytoskeleton is tubulin, which I mentioned earlier in connection with symmetries. As I said there, this remarkable molecule is a long tube composed of two units, alpha- and beta-tubulin, arranged like the black and white squares of a checkerboard. The tubulin molecule can grow by adding new units, or it can shrink by splitting backward from the tip, like a banana skin. It shrinks much faster than it grows, but both tendencies can be stimulated by suitable chemicals. The cell changes its structure by going fishing with tubulin rods in a biochemical sea. The rods themselves respond to the chemicals, which cause them to extend, collapse, or wave around. When the cell divides, it pulls itself apart on a tubulin web of its own creation.

Conventional fluid dynamics this is not. But it is undeniably *some* kind of dynamics. The cell's DNA may contain the instructions for making tubulin, but it doesn't contain the instructions for how tubulin should behave when it encounters a particular kind of chemical. That behavior is governed by the laws of chemistry—you can no more change it by writing new instructions in the DNA than you can write DNA instructions that cause an elephant to fly by flapping its ears. What is the fluid-dynamics analogue for tubulin networks in a chemical sea? Nobody yet knows, but this is clearly a question for mathematics as well as for biology. The problem is not totally unprecedented: the dynamics of liquid crystals, a theory of the patterns formed by long molecules, is similarly puzzling. Cytoskeleton dynamics, however, is vastly more complicated, because the molecules can change their size or fall apart completely. A good dynamical theory of the cytoskeleton would be a major component of morphomatics, if only we had the foggiest idea how to understand the cytoskeleton mathematically. It seems unlikely that differential equations will be the right tool for such a task, so we need to invent whole new areas of mathematics, too.

A tall order. But that is how mathematics grew in the first place. When Newton wanted to understand planetary motion, there was no calculus, so he created it. Chaos theory didn't exist until mathematicians and scientists got interesed in that kind of question. Morphomatics doesn't exist today; but I believe that some of its bits and pieces do—dynamical systems, chaos, symmetry breaking, fractals, cellular automata, to name but a few.

It's time we started putting the bits together. Because only then will we truly begin to understand nature's numbers—along with nature's shapes, structures, behaviors, interactions, proceses, developments, metamorphoses, evolutions, revolutions . .

We may never get there. But it will be fun trying.

FURTHER READING

Chapter 1

Stewart, Ian, and Martin Golubitsky, *Fearful Symmetry* (Oxford: Blackwell, 1992).

Thompson, D'Arcy, *On Growth and Form*, 2 vols (Cambridge: Cambridge University Press, 1972).

Chapter 2

Dawkins, Richard, "The Eye in a Twinkling," *Nature*, 368 (1994): 690–691.

Kline, Morris, *Mathematics In Western Culture* (Oxford: Oxford University Press, 1953).

Nilsson, Daniel E., and Susanne Pelger, "A Pessimistic Estimate of the Time Required for an Eye to Evolve," *Proceedings of the Royal Society of London, B*, 256 (1994): 53–58.

Chapter 3

McLeish, John, *Number* (London: Bloomsbury, 1991).

Schmandt-Besserat, Denise, *From Counting to Cuneiform*, vol. 1 of *Before Writing* (Austin: University of Texas Press, 1992).

Stewart, Ian, *The Problems of Mathematics*, 2nd ed. (Oxford: Oxford University Press, 1992).

Chapter 4

Drake, Stillman, "The Role of Music in Galileo's Experiments," *Scientific American* (June 1975): 98–104.

Keynes, John Maynard, "Newton, the Man," in *The World of Mathematics*, Vol. 1, ed. James R. Newman (New York: Simon & Schuster, 1956): 277–285.

Stewart, Ian, "The Electronic Mathematician," *Analog* (January 1987): 73–79.

Westfall, Richard S., *Never at Rest: A Biography of Isaac Newton* (Cambridge: Cambridge University Press, 1980).

Chapter 5

Kline, Morris, *Mathematical Thought from Ancient to Modern Times* (New York: Oxford University Press, 1972).

Chapter 6

Cohen, Jack, and Ian Stewart, "Let *T* Equal Tiger . . . ," *New Scientist* (6 November 1993): 40–44.

Field, Michael J., and Martin Golubitsky, *Symmetry in Chaos* (Oxford: Oxford University Press, 1992).

Stewart, Ian, and Martin Golubitsky, *Fearful Symmetry* (Oxford: Blackwell, 1992).

Chapter 7

Buck, John, and Elisabeth Buck, "Synchronous Fireflies," *Scientific American* (May 1976): 74–85.

Gambaryan, P. P., *How Mammals Run: Anatomical Adaptations* (New York: Wiley, 1974).

Mirollo, Renato, and Steven Strogatz, "Synchronization of Pulse-Coupled Biological Oscillators," *SIAM Journal of Applied Mathematics*, 50 (1990): 1645–1662.

Smith, Hugh, "Synchronous Flashing of Fireflies," *Science*, 82 (1935): 51.

Stewart, Ian, and Martin Golubitsky, *Fearful Symmetry* (Oxford: Blackwell, 1992).

Strogatz, Steven, and Ian Stewart, "Coupled Oscillators and Biological Synchronization, *Scientific American* (December 1993): 102–109.

Chapter 8

Albert, David Z., "Bohm's Alternative to Quantum Mechanics," *Scientific American*, 270 (May 1994): 32–39.

Garfinkel, Alan, Mark L. Spano, William L. Ditto, and James N. Weiss, "Controlling Cardiac Chaos," *Science*, 257 (1992): 1230–1235.

Gleick, James, *Chaos: Making a New Science* (New York: Viking Penguin, 1987).

Shinbrot, Troy, Celso Grebogi, Edward Ott, and James A. Yorke, "Using Small Perturbations to Control Chaos," *Nature*, 363 (1993): 411–417.

Stewart, Ian, *Does God Play Dice?* (Oxford: Blackwell, 1989).

Chapter 9

Cohen, Jack, and Ian Stewart, *The Collapse of Chaos* (New York: Viking, 1994).

Douady, Stéphane, and Yves Couder, "Phyllotaxis as a Physical Self-Organized Growth Process," *Physical Review Letters*, 68 (1992): 2098–2101.

Penegrine, D. H. G. Shoker, and A. Symon, "The Bifurcation of Liquid Bridges," *Journal of Fluid Mechanics*, 212 (1990): 25–39.

X. D. Shi, Michael P. Brenner, and Sidney R. Nagel, "A Cascade Structure in a Drop Falling from a Faucet," *Science*, 265 (1994): 219–222.

Waldrop, M. Mitchell, *Complexity: The Emerging Science at the Edge of Order and Chaos* (New York: Simon & Schuster, 1992).

Wilson, Howard B., *Applications of Dynamical Systems in Ecology*, Ph.D. thesis, University of Warwick, 1993.

Epilogue

Cohen, Jack, and Ian Stewart, "Our Genes Aren't Us," *Discover* (April 1994): 78-83.

Goodwin, Brian, *How the Leopard Changed Its Spots* (London: Weidenfeld & Nicolson, 1994).

INDEX